CAD and GIS Integration

Integration

Edited by
Hassan A. Karimi • Burcu Akinci

CRC Press
Taylor & Francis Group
Boca Raton   London   New York

CRC Press is an imprint of the
Taylor & Francis Group, an **informa** business

AN AUERBACH BOOK

Auerbach Publications
Taylor & Francis Group
6000 Broken Sound Parkway NW, Suite 300
Boca Raton, FL 33487-2742

© 2010 by Taylor and Francis Group, LLC
Auerbach Publications is an imprint of Taylor & Francis Group, an Informa business

International Standard Book Number: 978-1-4200-6805-4 (Hardback)

Library of Congress Cataloging-in-Publication Data

CAD and GIS integration / editors, Hassan A. Karimi and Burcu Akinci.
    p. cm.
  Includes bibliographical references and index.
  ISBN 978-1-4200-6805-4 (alk. paper)
   1. Geographic information systems--Computer-aided design. 2.
Geography--Mathematical models--Computer-aided design. 3. Software compatibility.
I. Karimi, Hassan A. II. Akinci, Burcu.

G70.28.C33 2010
910.285--dc22
                                                    2009025801

Visit the Taylor & Francis Web site at
http://www.taylorandfrancis.com

and the Auerbach Web site at
http://www.auerbach-publications.com

Contents

Preface

The need for CAD/GIS integration arises in many applications. Integration of computer-aided design (CAD) and geospatial information systems (GIS) can reduce many inefficiencies and errors made during design, planning, and execution of a project. It can also streamline decision making during operations. Many tasks, in particular engineering tasks in design, construction, and asset management throughout the life-cycle of an infrastructure or a facility, require knowledge of and skills in several interrelated and yet disconnected task-specific software and tools. CAD and GIS were developed separately, with decades of standalone existence. They have traditionally been used in tandem for making crucial decisions in many applications. Problem solving through CAD and GIS requires their proper integration for efficient and effective processing of data and operations.

This book provides knowledge about existing and emerging methodologies, techniques, and technologies for integrating CAD and GIS. It provides background on CAD and GIS from their early development to current trends and future directions for integrating CAD and GIS. It covers both the breadth and depth of knowledge in this area.

Chapters 1 and 2 introduce CAD and GIS, respectively, providing historical overviews, current trends, and insights on future directions. Chapter 1, contributed by Omer Akin, covers the history of CAD, which was developed in the 1960s; describes its current status, with a specific focus on building information modeling; and discusses future directions in CAD to address emerging architecture, engineering, construction, and facility management challenges. Such challenges include building performance modeling, life-cycle modeling, sustainable design, operations and maintenance efficiency, and manufacturing, as well as supporting interoperability and seamless data exchange between different CAD systems.

Chapter 2 is contributed by Piyawan Kasemsuppakorn, Duangduen Roongpiboonsopit, and Hassan A. Karimi. It provides an overview of GIS, with an emphasis on fundamental concepts and theories, as well as current developments and emerging trends. The topics discussed in the chapter include fundamental models for geographic representation and spatial analyses, as well as recently developed distributed GIS techniques, Geo Web Services, Mobile GIS, and 3D GIS. The chapter ends with a discussion on future trends, highlighting the need for seamless data exchange among different types of GIS.

Chapter 3, contributed by Omer Akin, discusses the rationale and challenges associated with the integration of CAD and GIS. It highlights historical differences between CAD and GIS in terms of their ability to create detailed geometry, to represent different levels of details, to incorporate data-centric versus graphic-centric perspectives, and to support single versus multiple users. It also provides an overview of different emerging vendor-specific CAD/GIS integration approaches and solutions. It concludes with a detailed discussion of several rationales and challenges that need to be addressed, as well as several methodologies that are being developed to streamline CAD/GIS integration.

After these introductory chapters, Chapters 4 and 5 provide an overview of the standardization efforts in CAD and GIS. Chapter 4, contributed by Semiha Kiziltas, Fernanda Leite, Burcu Akinci, and Robert R. Lipman, examines interoperable methodologies and techniques in CAD. It highlights the cost of interoperability in the U.S. facilities industry, and reviews different data standards and specifications that are being developed. It covers some of the early data exchange efforts (such as DXF and IGES) that predominantly target exchanging geometric information. It also discusses in detail recent standards for product model data exchange (e.g., STEP, CIS/2, gbXML) and for semantically rich building information model exchange (e.g., IFC, IFD). The chapter compares these standards in terms of the phase of the project that they target, the semantics that they incorporate, the usages that they have, the file formats they support, and the ability for extension that they provide.

Chapter 5, contributed by Carl Reed, provides an overview of interoperable methodologies and techniques in CAD/GIS integration from the perspective of the Open Geospatial Consortium (OGC), which is the predominant standardization organization for GIS. The chapter discusses the growing interest and activities targeting interoperability of CAD and GIS. It specifically considers six activities within the OGC that are relevant to integration of GIS/CAD/Building Information Models (BIM): Open GIS Web Map Services (WMS) Interface Standard, OGC Web Map Context Standard, OGC Web Feature Service Interface Standard, OGC and ISO Geography Markup Language Encoding Standard, OGC CityGML Encoding Application Schema, and OGC KML 2.2 Encoding Standard. It concludes with a discussion on the Architecture, Engineering, Construction, Owner, and Operator (AECOO) Test Bed, as a way to demonstrate GIS/CAD/BIM integration in the near future.

Chapters 6, 7, 8, and 9 review different emerging approaches that are being developed both in academia and in industry to address CAD/GIS integration challenges. Chapter 6, contributed by Mahsa Ghafourian and Hassan A. Karimi, examines the functional requirements for CAD/GIS integration to support seamless indoor/outdoor navigation. This chapter discusses various requirements of indoor/outdoor navigation in terms of the data models and coordinate systems being used, as well as different scales that need to be supported in navigation. It also highlights various technologies used for localization in indoor and outdoor environments, as well as different modes of transportation and user preferences that need to be considered.

Chapter 7, contributed by Michael J. Casey and Sriharsha Vankadara, discusses the role of semantics in CAD/GIS integration and current perspectives on interoperability and integration, such as syntactic versus semantic integration, and the use of semantic Web languages and services. It describes several cases within facility and infrastructure planning and security engineering areas that highlight the need for supporting semantic integration. This chapter also assesses building information modeling as an approach for enabling semantic integration of CAD and GIS.

Chapter 8, contributed by Tamer El-Diraby and Hesham Osman, discusses utilization of ontological approaches to enable semantically rich integration of CAD and GIS. It provides foundational knowledge on ontologies and covers several methods for creating ontologies. It also describes in detail an ontology developed for infrastructure products and related concepts for the city of Toronto. This offers an example to demonstrate the role of ontologies in streamlining the consolidation of heterogeneous data from CAD and GIS.

Chapter 9 is contributed by Burcu Akinci, Hassan A. Karimi, Anu Pradhan, Cheng-Chien Wu, and Greg Fichtl. It discusses the utilization of semantic Web services as a way to enable interoperability between CAD and GIS. The chapter specifically considers a set of interoperability challenges associated with CAD and GIS platforms, and describes a Web service–based approach that can enable semantic interoperability between CAD and GIS. Such a semantic Web approach incorporates task decomposition, ontology identification, Web service discovery and matching, and service composition.

Authors

Hassan A. Karimi is director of the Geoinformatics Laboratory in the School of Information Sciences at the University of Pittsburgh. His current research interests include navigation, mobile computing, geospatial information systems, location-based services, computational geometry, spatial data structures and databases, and grid/distributed/parallel computing. Hassan is an associate editor of the *Journal of Location-Based Services* (publisher: Taylor & Francis), the editor of the *Handbook of Research on Geoinformatics* (publisher: IGI Global), a co-editor of *Telegeoinformatics: Location-Based Computing and Services* (publisher: Taylor & Francis), and a member of the advisory board of the *Journal of Computers, Environment and Urban Systems* (publisher: Elsevier).

Burcu Akinci is a professor of civil and environmental engineering at Carnegie Mellon University. Her current research interests include modeling and reasoning with information-rich histories of construction projects and facilities/infrastructure systems to support decision making during construction planning and execution, commissioning, and operations and management. In relation to this area of research, she leverages building information models and data that can be collected using emerging sensing and data capture technologies. Burcu has written more than 30 journal publications and 40 conference proceedings, and has advised more than 15 PhD students and 10 MS students. She serves as part of the editorial board for the *Journal of Computing in Civil Engineering, Automation in Construction*, and *Advanced Engineering Informatics*.

Contributors

Omer Akın
Carnegie Mellon University
Pittsburgh, Pennsylvania

Burcu Akinci
Carnegie Mellon University
Pittsburgh, Pennsylvania

Michael J. Casey
George Mason University
Fairfax, Virginia

Tamer El-Diraby
University of Toronto
Toronto, Canada

Greg Fichtl
University of Pittsburgh
Pittsburgh, Pennsylvania

Mahsa Ghafourian
University of Pittsburgh
Pittsburgh, Pennsylvania

Hassan A. Karimi
University of Pittsburgh
Pittsburgh, Pennsylvania

Piyawan Kasemsuppakorn
University of Pittsburgh
Pittsburgh, Pennsylvania

Semiha Kiziltas
Middle East Technical University
Ankara, Turkey

Fernanda Leite
Carnegie Mellon University
Pittsburgh, Pennsylvania

Robert R. Lipman
National Institute of Standards and
    Technology
Gaithersburg, Maryland

Hesham Osman
UMA|AECOM
Toronto, Canada

Anu Pradhan
Carnegie Mellon University
Pittsburgh, Pennsylvania

Carl Reed
Open Geospatial Consortium
Wayland, Massachusetts

Duangduen Roongpiboonsopit
University of Pittsburgh
Pittsburgh, Pennsylvania

Sriharsha Vankadara
George Mason University
Fairfax, Virginia

Cheng-Chien Wu
University of Pittsburgh
Pittsburgh, Pennsylvania

Chapter 1

Current Trends and Future Directions in CAD

Omer Akın[*]

Contents

1.1　CAD: What's in an Acronym?

CAD stands for *computer-aided design*. In its early days it stood for *computer-aided drafting*. Others have coined acronyms containing words derived from *computing* and some other application area in the building infrastructure sector that range from CAE (computer-aided engineering) to CAFM (computer-aided facilities management), with dozens of variations in between. Since the middle of the last century, design professionals have been favoring computational applications in unprecedented numbers.

---

[*] With a contributor from Khee Poh Lam.

1

Exploring the underpinnings of this attraction is a compelling journey, that is, if one wishes to venture from where CAD has originated toward its likely fortunes.

Computation in its most powerful form provides virtual models, which are abstractions of reality, or a complex combination of them; we could call these *virtual worlds*, but this term has already been taken by others. These models or worlds allow users to explore, describe, predict, and control surrogate realities in order to gain practice with, if not insight into, the phenomenon represented by them. In short, computation is a powerful tool of designers (engineers, architects, and the like) for a myriad of reasons related to the virtual nature of their applications.

Through their core vocation, design professionals are deeply interested in virtual applications. In fact, design is the act of creating representations, not realities. The reality follows only if and when a *build professional*, as opposed to a *design professional*, realizes it. In doing this, she converts what is virtually present in the *blueprint* of a building, a ship, a fabric, or a robot, into a *real* building, ship, fabric, or robot. Thus, the cooperation between the design professional and the virtual world of computing is one that is predestined through predilections of a virtual kind.

Over the course of the last few decades, dozens of such matches have resulted in the many well-established flavors of CAD. The pioneers of the field, naturally, have come from those professions in the "AEC (Architecture-Engineering-Construction) industry" category in Table 1.1. This is largely due to their unequivocal commitment to design and its abstract nature, which has at least a 500-year lineage, dating back to the academies of Renaissance Italy.

The emergence of CAD, as is the pattern in most other fields, came in spurts and stops, not unlike Thomas Kuhn's description of advances in our scientific knowledge in his seminal book, *The Structure of Scientific Revolutions* [1]. To put a cleaner framework around this, let us identify the epochs in CAD that have led to shifts in its disciplinary underpinnings. We propose three such paradigms: (1) the first CAD paradigm, from 1960 to 1979, (2) the second CAD paradigm, from 1980 to 1994, and (3) the third CAD paradigm, from 1995 to 2004. The shifts in these paradigms, or "revolutions" in Kuhn's terminology, are events that took place around the dates demarking the end of one period and the beginning of the other. These dates are specific, but approximated in order to generally match the middles and ends of decades. In turn, these marker events came into being due to many developments that took place in the preceding paradigm period.

The first paradigm is the inception of CAD. In this period, the basic digital functions and representations of the practice came into being. In this form, CAD was, by and large, derived from manual practices and paper-based interactions. The second paradigm marks the emergence of computing software and hardware that successfully exploited the potentialities in the building design realm. The new technology was no longer an experimental tool, but one that was here to stay. The third paradigm arose from the level of sophistication that greatly enhances building delivery operations, standardization, and ease of input and output.

Table 1.1 Professions and Sub-Fields of Computer-Assisted Design

CAD in Architecture-Engineering-Construction (AEC) Industry		
CAD sub-fields	Architecture	Railroad, bridge, tunnel design
	Architectural engineering	Water supply and hydraulic engineering
	Interior design	Storm drain, wastewater, sewer systems
	Interior architecture	Mapping and surveying
	Civil engineering infrastructure systems	Heating, ventilation, air-conditioning (HVAC)
	Construction management	Factory layout
	Roads and highways	Chemical plant design
CAD in Mechanical Engineering		
CAD sub-fields	Automotive—vehicles	Machinery
	Aerospace	Shipbuilding
	Consumer goods	Bio-mechanical systems
CAD in Electronic Design Automation and Electrical Engineering		
CAD sub-fields	Electronic and electrical (ECAD)	Power systems engineering and PS-CAD
	Digital circuit design	Power analytics
Other CAD professions	Manufacturing process design	Industrial design
	Software engineering	Apparel and textile design
	Landscape design	Fashion design
	Bio-medical engineering	Lighting design

Today, as a result of these developments, CAD, both as a medium and a tool, has become independent of the manual, legacy processes. Out of this, what we can call I-CAD, or "intelligent" CAD applications, are emerging. From the very beginning, computers have offered the promise of augmenting designers' cognitive capabilities through functionalities that resemble human intelligence. These would take the form of enabling building infrastructures (equipment, components, and systems) to behave in ways that adjust to changing environmental conditions, estimating cost and schedule information, or providing histories of processes and products. Today, we are at the brink of realizing these non-trivial ambitions.

1.2 First CAD Paradigm: Emergence of CAD (1960–79)

The CAD revolution began in earnest at MIT in the mid-1960s. By today's standards, there was hardly enough going on in the AEC world at that time to call this a revolution; however two seminal works emerged from MIT: one by Ivan Sutherland [2] and the other by Steven Fenves and his coauthors [3], which were nothing short of spectacular. In these works, one understood how models of designs could be created in the computer, harnessing not only the logic of geometry of objects, but also their structural properties with precision and specificity. The virtual world of building design made a momentous entry into the digital world of computing.

Obviously, both Sutherland's and Fenves' work did not materialize out of thin air. The earliest computer graphics work dates back to the mid-1950s, with the SAGE (semi-automatic ground environment) system that was used in air defense command and control tasks, and the Whirlwind CRT (cathode-ray tube) display console at MIT's computer control room. These systems were built by Digital Equipment Corporation (DEC), which also built the TX1 and the DEC338 with "intelligent" graphics capabilities (Teicholz [4]). Around the same time, the surface patch technique critical for graphics work and the DAC/1 (design augmented by computer) system were being developed by Coons at MIT [5], and by Don Hart and Ed Jacks at GM (Kalay [6]), respectively.

Sutherland is responsible for the invention of the first CAD (computer-aided-drafting) system called *Sketchpad* [2]. With its devices of interactivity, and parametric and inheritance-based design concepts, Sketchpad was, and still is, nothing short of remarkable. It became a precursor to human computer interaction (HCI), parametric design, and object-oriented (OO) programming, advances which are still alive and well in boths the realms of CAD research and practice.

Sketchpad allowed data input through a tablet and electronic pen device, which raised important issues of tactile interaction (Figure 1.1), visual representation, spatial reasoning, and general human computer interactivity in CAD. Spatial operations with geometric entities like arcs, segments, and their orientations could be represented and manipulated on the CRT (cathode-ray tube) monitors while maintaining their parametric properties. These entities had schematic attributes, which become the industry standard in the software engineering field through object-oriented (OO) applications. Sketchpad also contributed to the development of technology that is commonplace today, like zooming, rotating, and set operations. Sutherland, throughout his long and illustrious career, directly or indirectly influenced many innovations including virtual reality, anti-aliasing techniques, shading algorithms, cognition-based programming, and 3D computing and printing languages.

Figure 1.1 Ivan Sutherland and his Sketchpad system (from http://www. cc.gatech.edu/classes/cs6751_97_fall/projects/abowd_team/ivan/ivan.html).

Around the same time at MIT, Steve Fenves made similar strides in the realm of structural design of buildings [3]. His application called STRESS (Structural Engineering Systems Solver) was the first and the founding system to execute structural analysis of a given structural system and assist the engineer to size and specify structural components.

From the beginning of his career, Fenves has been a pioneer in applying computer methods to engineering practice. After having completed his PhD on "experimental structural dynamics" in 1961, he accepted a post at MIT with Professor Charles Miller during 1962–63. Miller, a pioneer in his own right, had recently completed COGO, a general problem-solving capability for the domain of surveying. The initial concept for STRESS was to accomplish the same in the field of structural design, "invoked by terms that a professional in the field would use in giving instructions to a colleague" (Fenves [7]). The original agenda for STRESS included the following:

1. A problem-oriented textual input language patterned after COGO
2. Flexibility in problem size, so that the solution of small problems would not be penalized by the program's capability of solving very large problems
3. Complete generality in handling various framed structure types, e.g., frames or trusses
4. Complete generality of methods

With the guidance of Professor Miller and collaboration with cohorts such as Dr. Frank H. Branin, Jr., of IBM and several MIT researchers, including Dr. Bob Logcher; Dr. Sam P. Mauch; Ken Reinschmidt, an IBM graduate fellow; and graduate student Leon Wang, Fenves completed his work in short order. The first three objectives listed above were satisfied by 1964, creating a data description language that provided individually executable commands, a capability beyond a command language application.

Quickly recognized as the foundation of CAE (computer-aided engineering), STRESS became a standard application in the IBM-1120, a machine marketed to engineers at the time. Thus, a first generation of structural engineers was introduced to matrix structural analysis.

Fenves' many other contributions to the field include decision tables, knowledge-based representations for codes and standards, engineering information management, and Artificial Intelligence based methods for conceptual engineering design (Garrett [8]).

New ideas and prototypes followed these early CAD innovations with little or no real impact on the field of building design. In the early days, CAD was bogged down, among other reasons, for lack of easy access to computing, unavailability of widespread networking, obstacles to interoperability, and difficulty of input and output. Yet, the field of CAD made up for what was not readily available in computational hardware and software with enthusiastic trail blazing. Entrepreneurship by a handful of schools and companies, well documented in *CAD/CAM Handbook* (Teicholz [4]) and Kemper's *Pioneers of CAD in Architecture* [9], created the basic functionality of design by computer and research into CAD. Teicholz cites a dozen or so companies that helped the development of computer-aided graphics during the 1960s and 1970s (Table 1.2).

Table 1.2   Computer-Aided Graphics Companies of the Early 1970s and 1980s

Year	Company Name
1963	Adage
1968	Calma (later became part of GE)
1968	Evans and Sutherland
1968	Houston Instruments (later became part of Bausch and Lomb)
1968	Imlac
1969	Applicon (later became part of Schlumberger)
1969	Computervision
1969	Vector General
1969	Zeta Research (later became part of Nicolet-Zeta)
1970	Lundy
1971	Ramtec
1972	Megatec (later became part of United Telecommunications)
1972	Summagraphics
1974	Talos Systems (later purchased by Sander)

1.3 Second CAD Paradigm: Emergence of Specialized Computing (1980–1994)

This early interest in computer graphics translated into a remarkably robust CAD industry tapping into the resources of the lucrative construction industry, which happened to be generally successful during the 1970s and the 1980s. Even during the rollercoaster 1970s, energy conservation measures added fuel to the growth of CAD through performance simulation and measurement applications, which remain to this day a large component of digital design application.

Machover Associates Corporation's survey indicates that there were about one million engineers and 300,000 drafters in the U.S., 260 U.S. manufacturing companies with larger than $1 billion volume of sales per year, 750 U.S. manufacturing companies with sales between $100 million and $1 billion per year, and 30,000 U.S. manufacturing companies with sales between $5 million and $100 million per year (Teicholz [4]). Based on this data, we can chart a healthy growth picture for the CAD-CAM industry during the early 1980s (Figure 1.2).

Many corporations that developed their own in-house CAD software, such as Daussault, Lockheed, McDonnell Douglas, and Northrop, realized the potential of this market and began to vend CAD products. The estimates of worldwide growth of CAD-CAM, developed by Thomas P. Kurlak of Merrill Lynch in 1982 (Teicholz [4]), indicate a reliable growth picture (Table 1.3).

Several software platforms dominated the field in the 1970s and 1980s, including AutoCAD, ARIS, Micro Station, and Form-Z in the U.S., and Nemetschek Systems in Europe. Recently, this landscape has become much more complex with the entry of dozens of new platforms, including large 3D modeling and rendering software packages like MAYA and CATIA, and mainstream ones like SketchUP and TurboCAD, developed in the U.S., and ArchiCAD, developed in Europe.

The one unchanging feature in the CAD landscape from the beginning, particularly in the U.S., has been AutoCAD. From the outset AutoCAD has had

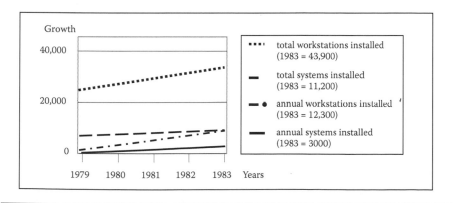

Figure 1.2   **CAD/CAM systems growth (1979–83).**

Table 1.3 Estimated Worldwide CAD-CAM Market by Application ($ millions)

	1979 (% growth)	1980 (% growth)	1981 (% growth)	1982 (% growth)	1983 (% growth)	Total	% of market
Mechanical	128 (+85)	225 (+75)	360 (+60)	443 (+23)	545 (+23)	1700	38
Electrical	98 (+84)	167 (+66)	223 (+40)	300 (+29)	410 (+37)	1210	27
A-E-C	58 (+100)	87 (+50)	138 (+59)	200 (+46)	302 (+50)	785	18
Mapping	32 (+100)	73 (+128)	111 (+52)	144 (+30)	180 (+25)	540	12
Other	18 (+82)	20 (+11)	30 (+50)	63 (+110)	83 (+32)	215	5
Total	334 (+87)	572 (+71)	872 (+52)	1150 (+30)	1520 (+32)	4450	

unequivocal dominance of the field due to customer loyalty. This has been attributed to its early entry and pivotal role in market dominance. Even during the early days when very few professionals used, computers or even considered computing as a positive influence on the field, Autodesk, the company conducting AutoCAD's research-and-development and marketing business, maintained this superiority. As a consequence of its willingness to develop new features in order to maintain a constant growth in step with the marketplace, Autodesk's data formats, DXF and DWG, have enabled it to exercise a firm hold on the CAD market.

By default rather than by design, Autodesk data formats have served as the "poor man's interoperability mechanism," linking up with other CAD software, including the competition that conformed to the format in order to hold onto a sufficient share of the market to survive. This data conformance also opened the door to interoperability between fields like structural, mechanical, electrical, and plumbing consultancies. In order to be competitive, other software platforms were left to their own ingenuity to invent significant features that were not existent in AutoCAD. Among these features there is ease of usability, appeal to early stages of design, high-end rendering, and data representation at the operations end of the design delivery process.

Since the early days in the 1950s, computer hardware capacity also has been improving. In fact, improvements taking place by orders of magnitude have resulted in the escalation of the rate of increase in terms of both RAM and CPU i.e., storage and processing memory. According to Moore's Law, every two years we are able to place twice as many transistors on the same integrated circuit that can work twice as fast because the electrons have half the distance to travel (Figure 1.3). Today a $1,000 computer's hardware capacity is the equivalent of the "computing power" of a creature between an insect and a mouse. At this rate of development, we are expected to catch up to the processing capacity of the human brain before 2020 (Kurzweil [10]).

With the maturation of more general computational capabilities, researchers and market forces enjoyed the emergence of digital prowess through unfathomable

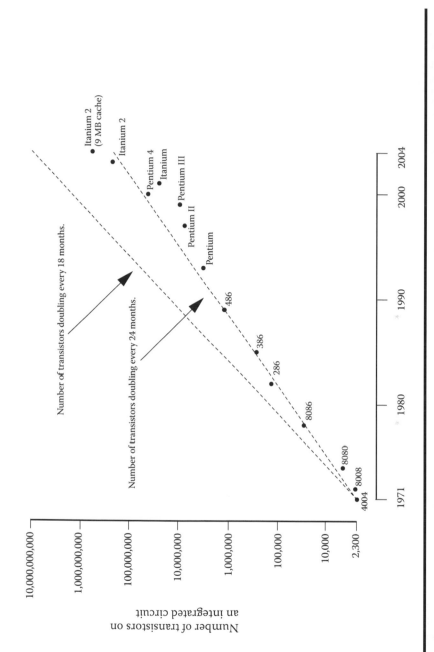

Figure 1.3 Growth of transistor counts for Intel processors and Moore's Law (logarithmic scale) (from http://en.wikipedia.org/wiki/Image:Moores_law.svg).

speed and capacity, greater sophistication in HCI, object-oriented (OO) software engineering applications, sophisticated graphics algorithms, and the World Wide Web (WWW). Finally, the field of CAD began to transition into a new paradigm commensurate with the ambitions of its early promise, through products that are ubiquitous, usable, powerful, interoperable, and smart.

Progress in theoretical and software terms also kept pace with these hardware advances. There were in pace with Nielsen revolutionized the field of usability design with his text on the principles of software usability, *Usability Engineering*, in 1994. Akın's [11] pioneering work on the *Psychology of Architectural Design*, and that of many others who subsequently contributed to this area, described critical benchmarks for researching task-based cognition in AEC fields. Interactive and immersive systems popped-up in the literature for all conceivable applications, including facility maintenance and management and operations (Lee et al. [12]), as well as "immersive audio-visualization of an artificial immune system" (Bentley et al. [13]). Eventually, these advances in software and hardware systems eventually ushered in the paradigm shift to information-based computing in the AEC domain.

1.4 Third CAD Paradigm: Large Data Models and Building Information systems (BIM) (1995–2004)

During this period, the interest in CAD among design professionals has been helped immensely by the larger graphics-oriented software applications like MAYA and CATIA. This trend created a near explosion of visualization efforts, which have been effective in "seducing" clients, as well as providing objectified representation of designs through solid modeling and rapid prototyping. The input/output functions of earlier computers were adapted to provide sophisticated modeling and manufacturing functionalities that overcame difficulties of manual modeling. Solid modeling capabilities were used to detect spatial conflicts and design errors.

In parallel to these developments, HCI became an interdisciplinary field bridging all of the professions and their sub-fields, such as those shown in Table 1.1 and others. It continued to building upon the early achievements of quantitative and qualitative models of task performance on the computer (Card et al. [14]). Furthermore, we observed the emerging theory of HCI ranging into as diverse areas as usability design (Nielsen [15]), cognition of task performance (Akin [11]), wearable computers (Pentland [16]), interactive-immersive systems (Malkawi [17]), and the social psychology of computer use (Turkle [18]).

Up until this point in time, design practice did not go through any basic changes attributable to computing, nor vice versa. While computational functions in other broadly used domains, like word processing, and spreadsheet applications, introduced fundamental changes in application areas as diverse as clerical work, accounting, administrative management, banking, payroll, commerce, medicine, and law, among many others, all but revolutionized the operations in these domains; building design behavior still conformed to the age-old patterns of document-based interaction as design and approval phases followed the conventional delivery process (Figure 1.4).

Recent developments in the area of building construction, which from time to time lunges forward with the earlier accomplishments of CAD research, have focused many of these ambitions upon one key concept: organizing all of the information about buildings in a multi-dimensional space. This space contains information on topics as diverse as life-cycle of building delivery, specialized building systems consultants, product and process coordination, cost-schedule-quality measures, and so on. Subsuming a space that is governed by shared product and process representations conforming to broadly accepted standards, this ambition can be summed up in one key concept: BIM or *building information systems* (Eastman et al. [19]).

During the 1960s through the 1990s, the field of CAD has been simply the conjunctive idea between building design and computing. But now, a fundamental shift is afoot that supports the view of building information as one that is focused neither on computation nor on the building domain alone. The world of BIM aspires to be one squarely focused on the *information* aspect of buildings with its own rules, practices, standards, interactions, and delivery cycles divorced, once and for all, from the idiosyncrasies of either area alone. This is a transformational impact on the way buildings are designed, delivered, and used.

There are three primary forces contributing to this movement: large data management, intra-task collaboration, and smart representations. First, the development of very large building models each containing hundreds of thousands, even millions, of distinct elements created through the modeling tools of early CAD led to the awareness that there are significant issues of data persistence, accuracy, and most importantly interoperability. Professionals initially using modeling software to facilitate documentation and ease of updates became rapidly aware of difficulties of interoperability and data exchange, laterally with consultants, and longitudinally along the time line. This marked the beginning of the shift of the paradigm from CAD to BIM.

Second, many of the information-based operations supporting the core function of design, such as programming, requirement management, commissioning, performance verification, operations and maintenance, costing, scheduling, quality control, building economics, and design value assessment, stand to benefit from BIM. When available, persistent, accurate, and computable data can have a radical effect on all of these operations. Use of data histories, interoperable mapping of information between tasks and phases, and just-in-time access to information can cut errors, cost, and time of completion dramatically.

Third, BIM addresses the need to embed intelligence in data such that product models are made up of modular units that "understand" their own scope of parts and features. The modular units allow values to be assigned to the parameters of these parts and features, and can transform their values by interacting with other units. These

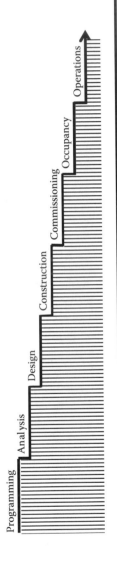

Figure 1.4 The conventional design delivery process: the waterfall model (Akin, 2004).

are some of the dimensions of intelligence that are being harnessed in BIM. In other words, the OO-modeling that has been around for a few decades now finds its just place in the scheme of things, providing persistent, ontological, and data dependent structures in the building information realm. Standards for interoperability, which come in many shades but with one dominant color: Industry Foundation Classes (IFC) provide the platform for data exchange intelligence so essential to BIM.

At long last, our focus of attention is being fixed on the nature of design delivery in the new epoch, rather than timesharing between two distinct concepts, namely design and computers. As is the case with all major technological innovations, initially the tools, concepts, and practices we use belong to an earlier paradigm, while new technologies merely imitate these older forms (Bhavnani, et al, [20]). Finally, the digital world of building delivery has arrived at the critical paradigm shift from an artificial conjunction of buildings and computation (CAD) to the transformative new concept of buildings as information (BIM).

1.5 The Future of CAD: 2005 and On

The fundamental concepts that make-up BIM are founded in the consideration of design as the management of information that exists at different stages of the formal building life-cycle and delivery process (Figure 1.4). Due to the emphasis on information and its management, the industry has shifted its focus to the representation of entities that are meaningful at the various stages of building delivery, such as physical components (walls, windows, HVAC units), their persistent and historic forms throughout the building life-cycle, and their transmission to interested parties (owners, consultants, regulatory agencies).

1.5.1 The Building Information Modeling Movement

The idea of BIM, neither novel nor new, is extremely timely. It has advanced quite rapidly past the emergence stage for a paradigm shift, and has become a fully blown movement. The context for its support is present and robust. In BIM, as is the case in all movements that enjoy unprecedented coalescence of support from diverse quarters, its *context* is the key to its persistent success (Gladwell [21]). This context is defined by some of the factors already described earlier: the need for management of large data repositories, intra-task collaboration, and smart representations. Existing CAD systems, such as Autodesk's REVIT, provide evidence to make a case for interoperability; not necessarily along horizontal lines crossing entire facility life-cycles, but at least along "vertically integrated business functions" (NIBS [22]).

The lack of lifecycle interoperability is probably the most acute productivity issue that the AEC sector faces. In the current manual process, there are repeated information buildup efforts required at the beginning of each phase, and major information loss at the end. By and large, this can be avoided in the "interoperable data process" modality (Figure 1.5). This point underscores both the promise and

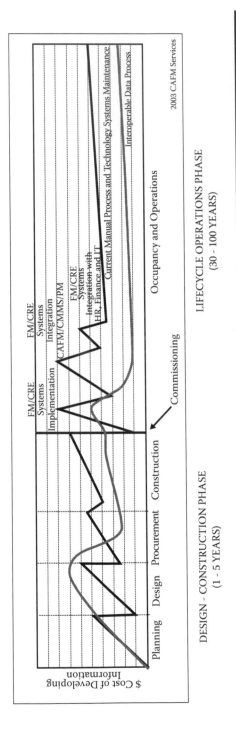

Figure 1.5 **The pay-now-or-pay-later chart for the AEC domain (Courtesy of Andy Fuhrman, International Facility Management Association, 1 E. Greenway Plaza, Suite 1100, Houston, TX).**

the challenge the BIM movement currently faces. Even as we witness the publishing of a standard for BIM by NIBS-FIC, and the authoritative handbook on the subject by Eastman and his colleagues [19], the direction and prospects for the success of BIM are still being debated, even by those who are responsible for its inception.

Some claim that BIM exhibits the classical signs of what is currently wrong with the AEC sector: disjunctive and distributed data, practices, and "standards" (Tardif [23]). There appears to be little dispute about this position; but ideas about its potential remedies seem to be contested by experts. Should there be a standard BIM practice and ontology centrally maintained and controlled in the near future? How realistic is it to achieve such a centrally placed BIM system? Should we rather support, for the sake of realistic expectations, a coordinated but distributed BIM? Is this difficulty an artifact of the culture and turf of building construction, rather than digital technology? Are there existing organizations and systems that can become the natural stewards of the centralized BIM, like the *buildingSMART* alliance of NBIMS (Smith [24])? This intense debate around BIM has been fueled by several concurrent technologies, including the modeling software offered by CAD vendors; the process and product model standards offered by international agencies, such as Standard for the Exchange of Product Model Data (STEP), International Agency for Interoperability (IAI), and their third party contributors; and the communication networking enabling remote operations, through the Internet.

The emergence of BIM as a groundbreaking technology and a common cause for those involved in the industry, even from the beginning, is the culmination of many large and small efforts distributed throughout the industry and academic institutions. The call to arms of this movement, however, was led by the National Institute of Building Standards on February 20, 2006:

> The National Institute of Building Standards (NIBS) through its Facility Information Council (FIC) has formed a committee to create the National Building Information Model Standard (NBIMS). The Standard is considered to be a critical element in reforming business practices in the capital facilities industry and recapturing at least $15B annually lost due to inefficiencies. The NBIMS team is actively soliciting participation from public and private organizations and individuals and, to date, 26 organizations representing most aspects of the facility life-cycle and many types of associated businesses have signed the charter.

This announcement went on to emphasize the importance of this effort for "reforms in industrial manufacturing and supply chain management" and "the long list of stakeholders; which include owners, architects, real property brokers, financiers, lawyers, insurers, engineers, materials scientists, manufacturers, constructors, facility managers, occupants, and services vendors." It identified the goals of this effort as "providing a common language for describing facility information, common views of information based on the needs of businesses engaged in all aspects of facility commerce, and common standards for sharing data between businesses and their data processing applications."

Stated in these most broad terms, it is difficult to distinguish the NBIMS organization and its goals from those of international standards organizations like STEP or IAI. However, the final product of the NBIMS committee, released in 2007, more clearly indicates the scope of its work and its objectives. In response to the lack of data interoperability along the lifecycle of a facility, NBIMS 2007 addresses standards for *Early Design, Construction to Operations Building Information Exchange (COBIE), Business Process Roadmaps,* and other guidelines and support documents. The NBIMS document states: "Reference standards in the NBIM Standard provide the underlying computer-independent definitions of those entities, properties, relationships, and categorizations critical to express the rich language of the building industry." It also references IFC by STEP and *OmniClass** as its data standards. It proposes to measure the BIM application in a given context via the Capability Maturity Model widely used to validate the quality of software applications in the 1980s and 1990s (Chrissis [25]).

The *BIM Handbook* by Eastman and colleagues provides good coverage of the tools that are available and their interoperability and data standardization ranges. It also considers the future of BIM and how it can help transform this effort in the AEC industry. This is a fast moving target and it is still being formed by the market forces as well as three decades of research, started, interestingly enough, by Eastman and Fenves back in the early 1970s at Carnegie Mellon University.

In an effort to implement the concepts underlying BIM in the CAD marketplace, many software vendors are taking preliminary but serious steps toward OO-modeling and design data exchange. The challenges that still remain include the seamless extension of these capabilities to tasks outside of the core architectural ones, such as mechanical, structural, electrical, egress, emergency, and cladding systems, among others. Further extensions are needed to cover other design delivery stages, including requirements engineering and performance commissioning. While progress is being made here, there still remains a great deal to be done. The challenges are in the volatility of product and process information, lack of standards and standardization, lack of interoperability protocols, and industry's slow acceptance of digital data standards and exchange strategies.

---

* The OmniClass Construction Classification System (known as OmniClass™ or OCCS) is a new classification system for the construction industry. OmniClass is useful for many applications, from organizing library materials, product literature, and project information, to providing a classification structure for electronic databases. It incorporates other extant systems currently in use as the basis of many of its tables—MasterFormat™ for work results, UniFormat for elements, and EPIC (Electronic Product Information Cooperation) for structuring products (from http://www.omniclass.org/).

1.5.2 Building Performance Modeling

Khee Poh Lam

Advancements in building performance simulation over the past two decades have been significant, with new and improved computational tools that address the changing needs of architectural design throughout the building delivery lifecycle (Figure 1.4). The ultimate goal of these computational developments is to support sustainable architecture and the creation of healthy, comfortable, and productive habitats for human activities. Ironically, defining such human activities accurately as input factors in performance modeling remains probably the single most complex and challenging task.

Energy-based simulation tools continue to evolve to address, among others, two major objectives: first, to make simulation tools more accessible to the architectural profession; and second to enable effective "real-time" sharing of design information between the members of the entire project design team. Accessibility entails offering cost effective, ubiquitous (Web-based) simulation services to architects/engineers, especially in the early design phase. Green Building Studio (GBS) is one such example. The service provides quick preliminary energy prediction, based on the architect's design, represented by an OO model, and specifying two basic input parameters: building type and geographical location. Based on these, the service derives a set of assumed building specifications (construction type, mechanical system, etc.) from its increasingly rich database of buildings in various cities in the United States. It then uses the well-known DOE-2 energy simulation engine to generate the results. GBS also provides the input file, which can be used for further detailed design application and analysis by engineers.

Underlying this simulation service is the essential capability of connecting the geometric CAD model and the energy computation model. Such data "interoperability" has been the focus of organizations such as the IAI and the Green Building XML Schema [27]. While their implementation strategy may differ (top-down versus bottom-up), their missions are similar: to define and publish specification schemas for BIM, as a basis for sharing building information globally throughout a project lifecycle, across professional disciplines and computational design-support applications.

Besides exploring new theories in computational design support systems to meet changing needs in an information technology world, efforts in this area are deployed to create "seamless" and "intelligent" interfaces between CAD and performance simulation engines. For instance, in relation to the Leadership in Energy and Environmental Design (LEED) rating system, Yudelson [28] in "The Change Function" observed that "those in a position to determine the future of LEED should continue to reflect on how to make it more transparent and user-friendly to those in the trenches of building design, construction, and operations."

Recent research work also involves creating a "seamless" interface between the most prominent BIM of our day, REVIT, the CAD modeler, and the Radiance Lighting Simulation (RLS) engine to support early design exploration in lighting performance. The tool imports a REVIT model and, based on the specified building type and location, the necessary Radiance input files are generated and populated with "appropriately assumed" relevant building input data (where data is missing) for that project. Furthermore, the tool also automatically analyzes the building spatial configuration as well as the lighting results to determine whether the design fulfills the requirements of the LEED indoor Environmental Quality (EQ) Credit 8.1 (Daylight 75% of spaces) and Credit 8.2 (Views for 90% of spaces).

The World Business Council for Sustainable Development recently published their first report on Energy Efficiency in Buildings. It states that buildings are responsible for at least 40% of energy use in many countries, mostly by consuming energy derived from fossil fuels. Worldwide energy consumption by buildings is expected to grow 45% over the next 20 years. The building industry is being challenged to create energy efficient and high-performance buildings, which starts right at the project inception, when the client meets the architect for the first time to formulate the design brief. Cost-effective and sophisticated (but user-friendly) tools are constantly being developed and offered to the multi-disciplinary design team to concurrently support these complex simulation tasks. A new generation of designers is being trained with knowledge of these tools and their application in real-world conditions.

1.5.3 *Other Allied Developments*

While the areas of application covered in the above sections signify the core developments in the domain of CAD, there are several important developments which we have not covered simply because their lineage into the foundations of CAD are not as deep. However, these areas are important for the future of CAD and BIM and they require at least a brief introduction here: intelligent graphics, architectural robotics, requirements modeling, performance evaluation, requirement specification, construction site modeling, building commissioning, operations maintenance and management, and economic modeling.

Graphics has been a bulwark of early computational design. In fact it has been almost a synonym for CAD starting with Sutherland's work in the 1960s. Today, computer-aided graphics has yielded its position to approaches that exploit the intelligence programmable into computers rather than algorithmic representations of Euclidian geometry. These include parametric geometry (Moustapha et al. [29]), ontological languages for geometric entities (Stouffs et al. [30]), virtual and augmented reality applications (Chan [31]), and sketch understanding systems (Do [32]).

One of the intellectual extensions of the work in the graphic realm into pioneering applications can be found in the physical manifestation of design through digital technology. This includes digital fabrication technologies at one end, and architectural robotics applications at the other. Digital fabrication involves the rapid prototyping and modeling of designed entities in physical model form that can assist both client interactions and *in situ,* permanent installation of the designed object. Architectural robotics on the other hand makes operable parts for architectural settings that fulfill functions that heretofore have been unavailable in the market place of architecture, such as self-configuring walls, energy use feedback instruments, and the like (Gross, et al, [33])

The issues raised by lifecycle modeling present challenges and opportunities to integrate the graphic-visual aspects of design with its non-graphic aspects, namely design requirement specification and modeling (Ozkaya et al. [34]). Integration of requirement management and design is long overdue, since many of the lifecycle decisions in facility maintenance and operations can benefit from the availability of design intent, the key concept captured in requirements. Building commissioning is one of these application areas that is rising in prominence (Turkaslan-Bulbul et al. [35]). It provides one of the most serious opportunities we have for evaluating buildings and their validity to address client needs.

One of the critical areas of CAD emergence in the AEC sector is in the area of modeling for construction site management and planning (Akıncı et al. [36]). This is one of the most critical phases of building delivery that potentially impacts their cost, schedule, and quality, the three key performance areas of the AEC sector. There is a plethora of application areas including construction process planning, modeling, prediction, control through the use of remote sensing, data encoding, laser technology, simulation, solid modeling, and the like.

Once the building is turned over to the owner, its operation, maintenance, and management challenges begin in earnest (Lee et al. [12]). Often from the first day there are insurmountable issues in setup, calibration, diagnosis, repair, alteration, as-built records, user requests, inventory record keeping, emergency intervention, routine maintenance and the like. Innovative tools and approaches for many of these are being developed using digital tools to simulate, model, and otherwise provide just-in-time information to fieldworkers in this multi-billion-dollar segment of the U.S. economy.

Finally, there is a great need to develop intelligent assistance for economic decision making in all facets of the building delivery process. New economic models are emerging that take into account not only the traditional economic modeling issues like discount rates, net present value analysis, and value engineering, but also the real estate and design added-value analyses (Akın [37]). Integrating these methods of forecasting cost and budget with BIM technologies will no doubt provide significant functionalities in the AEC management domain.

In closing, it is worthwhile to note that with all of the challenges that the complex world of the building delivery process offers, and the rapidly developing digital capabilities that range from innovative input output of data, to actuation of physical objects, to innovative software applications that emulate human intelligence, the future of CAD, or shall we say BIM, is bright indeed. While the challenges, particularly the current ones we are facing in terms of manufacturability, standardization, and interoperability, are extremely tough, so are the ingenuity and technological prowess of those in the academic and business worlds aimed at addressing these challenges.

References

[1] Kuhn, T. (1970) *The Structure of Scientific Revolutions.* The University of Chicago Press, Chicago, IL.

[2] Sutherland, I. (1963) "Sketchpad: A Man Machine Graphical Communication System," PhD dissertation, MIT, Cambridge, MA.

[3] Fenves, S.J., Logcher, R.D., Mauch, S.P., and Reinschmidt, K.F. (1964) *STRESS: A User's Manual—A Problem-Oriented Computer Language for Structural Engineering,* MIT Press, Cambridge, MA; and Fenves, S. J. *Problem-Oriented Languages for Man-Machine Communication in Engineering,* pp. 43–56.

[4] Teicholz, E. (1985) *CAD-CAM Handbook.* McGraw-Hill Book Co., New York.

[5] Coons, S. A., "An Outline of the Requirements for a Computer-Aided Design System," MIT Report ESL-TM-169, March 1963.

[6] Kalay, Y.E. (2004) *Architecture's New Media: Principles, Theories, and Methods of Computer-Aided Design.* MIT Press, London.

[7] Fenves S. J. (2006) "What Makes and Doesn't Make a 'Killer App,'" in *Civil Engineering: A Retrospective Evaluation,* Maribor Talk.

[8] Garrett, J. (2001) "Special Issue on Computer Aided Engineering in Honor of Professor Steven J. Fenves," *Journal of Engineering with Computers,* 17, 2, pp. 93–94.

[9] Kemper, A. M. (1985) *Pioneers of CAD in Architecture,* Hurland/Swenson, Pacifica, CA.

[10] Kurzweil, R. (2003) "The Future of Intelligent Technology and Its Impact on Disabilities," *Journal of Verbal Interaction and Blindness,* 97, 10, October.

[11] Akın, O. (1989) *Psychology of Architectural Design,* Pion, Ltd, London.

[12] Lee, S. and Akin, O. (2008) "Shadowing Tradespersons: Inefficiency in Maintenance Field Work" in print *Automation in Construction.*

[13] Bentley P.J., Novakovic G., and Ruto A. (2005) "Fugue: An Interactive Immersive Audiovisualization and Artwork Using an Artificial Immune System," Department of Computer Science, University College London, London.

[14] Card, S. K., Moran, T.P., and Newell, A. (1983) *The Psychology of Human-Computer Interaction* Lawrence Erlbaum Associates, Hillsdale, NJ.

[15] Nielsen, J. (1993). *Usability Engineering,* Academic Press: San Diego.

[16] Pentland, A., "Wearable Computers," *IEEE Micro,* vol. 19, no. 6, pp. 9–11, Nov/Dec. (1999).

[17] Malkawi, A., Srinivasan, R., Jackson, B., Yun Y., Kin C., and Angelov, S. (2004) "Interactive, Immersive Visualization for Indoor Environments: Use of Augmented Reality, Human-Computer Interaction and Building Simulation" in proceedings of the *Eighth International Conference on Information Visualisation,* 14–16 July, pp. 833–838.

[18] Turkle, S. (1984) *The Second Self: Computers and the Human Spirit,* Simon and Schuster, New York.

[19] Eastman C., P. Teicholz, R. Sacks, K. Liston (2008) *Handbook: A Guide to Building Information Modeling for Owners, Managers, Designers, Engineers and Contractors.* John Wiley and Sons Inc., Hoboken, NJ.

[20] Bhavnani, S. K., John, B. E., and Flemming, U. (1999) "The Strategic Use of CAD: An Empirically Inspired, Theory-Based Course." *Proceedings of the SIGCHI Conference on Human Factors in Computing Systems,* Pittsburgh, PA, pp. 183–190.

[21] Gladwell, M. (2000) *The Tipping Point: How Little Things Can Make a Big Difference.* Little Brown, London.

[22] *NIBS* (2007) *National Building Information Modeling Standard (NBIMS) Version 1.0 Part 1 Overview, Principles and Methodologies.* National Institute of Building Sciences (NIBS) and Facilities Information Council (FIC), Washington, D.C.

[23] Tardif, M. "BIM Implementation: Applying Lessons Learned," featured article, *AEC Café,* (http://www10.aeccafe.com/nbc/articles/view_article.php?section=CorpNews&articleid=405748).

[24] Smith, D. (2007) "An Introduction to Building Information Modeling (BIM)." *Journal of Building Information Modeling* Fall, pp. 12–14.

[25] Chrissis, M.B., Konrad, M., and Shrum, S. (2003). *CMMI: Guidelines for Process Integration and Product Improvement.* Addison-Wesley Professional, New York.

[26] Green Building Studio (http://www.greenbuildingstudio.com/gbsinc/gbs-web.aspx).

[27] Green Building XML Schema (http://www.gbxml.org/about.htm).

[28] Yudelson, J. (2006) "The Change Function." *Green Buildings,* July 31 (http://www.igreenbuild.com/cd_2503.aspx).

[29] Moustapha, H. and Akın, O. (2004) "Strategic Use of Representation in Architectural Massing," *Design Studies* 25, 1, 31–50.

[30] Stouffs, R. and Krishnamurti, R. (2006) "Algorithms for the Classification and Construction of the Boundary of Shapes." *Journal of Design Research* 5, 1, 54–95.

[31] Chan, C.S. (1994) "A Hypermedia Tutoring for Multimedia Tasks" in the *Second International Conference of Design and Decision Support Systems in Architecture and Urban Planning.* Vaals, The Netherlands, August 15–19.

[32] Do, E.Y-L. (2005) "Design Sketches and Sketch Design Tools." *Knowledge Based Systems* 18(8): 383–405.

[33] Gross, M.D. and Do, Y-L. (2007) "Environments for Creativity—A Lab for Making Things." *Proceedings, Creativity and Cognition 2007, ACM SIGCHI,* pp. 27–36.

[34] Özkaya, İ. and Akın, Ö. (2006) "Requirement-Driven Design: Assistance for Information Traceability in Design Computing" in *Design Studies,* 27, 3 (2006), pp. 381–398.

[35] Turkaslan-Bulbul, M.T. and Akin, O. (2005) "Computational Support for Building Evaluation: Embedded Commissioning Model" in *Proceedings of the First Conference on the Future of the AEC Industry,* Las Vegas, Nevada and in *Journal of Automation in Construction* 15(4): 438–447.

[36] Akinci, B., Pradhan A., and Kiziltas, S. (2008). "Analyses of Data Sources for Multi-Source Data Fusion to Support Project Management Tasks" *Proceedings of 2008 NSF Engineering Research and Innovation Conference,* Knoxville, Tennessee, January 7–10.

[37] Akın, O. (2006) *A Cartesian Approach to Design Rationality.* The METU Press, Ankara, Turkey.

Chapter 2

Current Trends and Future Directions in GIS

Piyawan Kasccaemsuppakorn, Duangduen Roongpiboonsopit, and Hassan A. Karimi

Contents

2.1 Introduction

Even though GIS, the acronym for geospatial information system, has been around since the 1960s, many people recognize GIS only as a mapping tool. This understanding is misleading, since GIS provides more features and capabilities than simply visualizing geospatial data on a map. The U.S. Geological Survey (USGS) defines a GIS as "a computer system capable of assembling, storing, manipulating, and displaying geographically referenced information or geospatial data" [1]. Although the predecessors of GIS in the field of cartography and mapping have a long history spanning hundreds of years, GIS is a relatively young field. The first GIS, Canada Geographic Information System, was developed and launched in the 1960s. In the 1970s–1980s, the GIS industry had substantial growth, and several world-leading companies in GIS software were established including Environmental Systems Research Institute (ESRI) and Intergraph. GIS has been evolving along with developments in computer technologies and has become an essential component of modern information technology. For a detailed history of GIS, refer to [2–4].

GIS, as a decision support system, can assist users in explaining and solving real-world problems in various fields such as land and resource management, surveying and mapping, urban planning, market analysis, geology, hydrology, and so on. In essence, the abilities of GIS are different from other traditional information systems in that it integrates database operations and statistical analysis with unique visualization on a map [5]. While some similarities between GIS and computer-aided design (CAD) are introduced in Chapter 1, many differences among them exist. The major difference between GIS and CAD is that GIS is generally used to model the real world in large geographic extents, whereas CAD is often used to design objects within a limited space that may not exist in the real world [6]. Combining information available in GIS and CAD calls for new methodologies and models that integrate real-world phenomena while maintaining the details of objects for facilitating computation and representation.

The purpose of this chapter is to provide readers with the fundamental concepts and theories, the current developments, and the future directions in GIS. The structure of this chapter is as follows: Section 2.2 provides an overview of the fundamentals of GIS including basic components and functions, geographic representation, spatial analysis, and practical applications. Section 2.3 discusses the current state-of-the-art GIS technologies and developments. Finally, perspectives on future directions of GIS are discussed in Section 2.4.

2.2 Fundamentals of GIS

In this section, we introduce the basic components and functions of GIS, then describe geographic representation followed by spatial analysis. Finally, examples of major practical applications are illustrated. The reader interested in more details is referred to references 7–11.

2.2.1 GIS Components and Functions

A GIS requires six key components to work together: network, hardware, software, data, people, and procedures. An overview of each component is described as follows.

First, a *computer network* is the infrastructure that interconnects two or more computers and other devices for resource sharing and parallel computation [12]. A computer networks includes local-area networks (LANs), wide-area networks (WANs), and the Internet. These networks are a fundamental component of GIS, with their ability to enhance the accessibility and reusability of geo-referenced data and analysis tools.

Second, *hardware in GIS* refers to physical components of a computer system that are composed of a CPU to run the software; disk storage to store large amounts of data and programs; input devices such as digitizers, scanners to convert data; and output devices to display the results [13]. At present, GIS applications not only operate on desktops but are also available through laptops, personal digital assistants (PDAs), in-vehicle devices, and cell phones.

Third, *GIS software* is responsible for generating, storing, analyzing, manipulating, and displaying geographical data. Currently, many different GIS software packages are available in both commercial (e.g., ESRI, Intergraph, and Autodesk) and open source (e.g., GRASS) products.

Fourth, *GIS data* includes both geospatial and non-geospatial data for representing selected geographical areas, objects, or phenomena to serve specific purposes. A GIS database consists of a digital representation of a selected geographical area. Developing a GIS database is one of the most time consuming and costly tasks in implementing GIS projects. This is mainly because collecting geospatial data from various existing maps, field observations, and sensors is laborious, and converting the collected data into a standard digital form is expensive.

The fifth key component of a GIS is *people*, who can be categorized into three classes based on their roles: viewers, general users, and GIS specialists [11]. The largest class of users consists of viewers, who only query and visualize referential materials from a geographic database for presentation purposes. General users employ basic GIS operations for supporting business decision and analysis, such as locating customer location, optimal routes for transporting goods or services, or tracking shipped products. GIS specialists are those who are trained in GIS, understand the underlying concepts of GIS, and provide technical supports to the other two classes of users.

Last, *procedures or methods* are required to ensure that GIS activities, such as how to retrieve, store, transform, and analyze spatial data, maintain high quality, and meet the needs of the organization. Generally, these procedures are used to describe the steps taken in GIS development and comply with an implementation plan and business rules of each organization.

In general, GIS functions are identified through four broad categories including resources inventory and data sharing, spatiotemporal information management, network analysis, and spatial analysis [14]. *Resources inventory and data sharing* is needed to combine and represent the required data from heterogeneous data sources into a uniform schematic view. *Spatiotemporal information management* incorporates the temporal, or time dimension, into geographic information because it is an important factor in the representation of a change in location, size, orientation, and the form of spatial objects, which improves decision making in many applications. *Network analysis* includes connectivity analysis, path finding, and proximity tracing, which are necessary for a variety of applications, such as a navigation and guidance system. *Spatial analysis*, one of the most important GIS functions, provides a broad range of powerful spatial data modeling and analysis using geometrical and topological properties.

2.2.2 Geographic Representation

Representation plays a crucial role in developing computational systems. Likewise, in GIS a set of physical features or phenomena, over, on, or under the surface of the Earth is represented. This representation is usually referred to as geographic representation. As real-world geographic phenomena are extremely complicated and rich in variety, it is impractical to capture infinite details and store within limited digital resources of computational systems. Thus, only the most important features that reflect the real-world phenomena are represented explicitly for solving or understanding a problem in GIS. However, while a set of features is represented, there are many possible choices, such as models, level of details and time period, that are available to GIS designers to best fit their needs. Geographic representation has an influence on interpolation and analysis [15] and "finding the appropriate representation can be a major part of a problem-solving effort" [16].

2.2.2.1 Conceptual Models of Space

Several abstract conceptual models with different aspects of space could be used to represent geographic phenomena. These conceptual models allow us to perceive a phenomenon in a certain way. The two fundamental abstract conceptual models of space, object-based model and field-based model, are widely used in the GIS community.

The object-based model views the world as an empty space occupied by discrete objects (or entities) that can be recognizable. Each object has its own attributes, boundaries, and location. Objects may be natural geographic phenomena (e.g., lakes, rivers, islands, mountains, forests), human-made phenomena (e.g., buildings, roads, utilities, administrative region), or living things (e.g., humans, animals). Although this model treats each individual phenomenon as an isolated object, these objects could have some relationships. For example, a lake is located inside the boundary of a national forest. A special property of viewing things as objects is that they are countable. Thus, we can perceive the characteristics of phenomena by using statistical analysis. Notably, while the object-based model is appropriate for phenomena that have well-defined boundaries, there are natural phenomena whose boundaries change over time. Furthermore, the precision of a model at one instant may be different at another.

The field-based model represents the geographic world by a number of variables or attributes in continuous Cartesian coordinates across some regions of space. The attributes can be measured at any point on the Earth and, naturally, their values are smooth and continuous across the space. Examples of phenomena that may be represented as fields are temperature, air pressure, elevation, concentration of pollutants, crop fields, or green areas. Fields may be represented in two, three, or four (if time is included) dimensions depending on applications. For example, a 2D elevation model, known as digital elevation model (DEM), has a single value of elevation at any given location in x and y coordinates. The field-based model is often adopted when there is insufficient information about precise boundaries of the phenomenon.

There is no exact criterion to select one model over the other. The choice of a model sometimes depends on the data available; for example, one would adopt the field-based model if the observed data are satellite imagery or adopt the object-based model if the input data are points collected by using a Global Positioning System (GPS) receiver. The choice of a model sometimes depends on the purpose of the analysis. For example, the field-based model might be appropriate if the expected outcome of an analysis of a terrain surface is continuous slopes. In addition, the choice of conceptual models depends on the technical knowledge of the designer.

2.2.2.2 Computer Data Representation Models

Even though a conceptual model allows us to view a phenomenon in a certain way, it is not designed to deal with digital representation in computers. The conceptual model may still contain an infinite number of details about a geographic phenomenon. Thus, to transform information from conceptual models to digital formats, two board categories of spatial data models, vector and raster, are popularly employed. Vector and raster can represent either the object-based or the field-based model; however, in practice, vector is closely related to the object-based model and raster is closely related to the field-based model.

Vector model represents phenomena as a series of basic geometric components or spatial primitives, which consists of *points, lines (arcs)*, and *areas (polygon)*. Each primitive is usually defined by a set of coordinates. In 2D models, a point can be specified by a pair of x and y coordinates. A line is usually described by a series of point coordinates. A *polyline* or path is a curved line composed of consecutive lines.

An area or polygon is typically defined by one or more closed polylines that form a boundary. Objects represented by the vector data model are often called features.

Vector model can be generally divided into two types of structures: *simple* and *topological*. The simple structure stores only individual locations of features without recording spatial relationships among them. This simple type is sometimes called "spaghetti," since primitives, especially lines, can overlap. Several methods can be used to store and manage this simple vector information, such as a list of coordinates spaghetti, and vertex dictionary [17]. Although storing simple features in a database is simple and easy to manage, it does not contain spatial relationships among features explicitly. Storing the relationships among features requires a large storage space due to numerous duplications of simple features. On the other hand, the topological structure stores both locations and spatial relationships of a set of simple features within a space without duplicating features. For example, if two polygons are adjacent sharing a boundary, the shared line representing the boundary will be stored only once and it is recorded as neighbors in the database.

Raster model represents the phenomena of a selected surface by dividing the space into rectangular cells, usually square cells known as *pixels*. This model is sometimes called *grid model* since it resembles a grid of cells. Only some cells contain spatial coordinates, while other cells within the same raster model can infer their location using the ordering of the matrix, unlike the vector model that explicitly stores its coordinates. Besides, each cell contains an attribute value associated with the cell's location. Thus, the variation of a continuously varying attribute can be represented by a sequence of cells. The cell size is determined by the varied resolution of the variation of an attribute. Low resolution indicates a large cell size while high resolution indicates fine grain cells. Mostly, raster data are obtained through information observed by remote-sensing technology (e.g., satellite imagery and aerial photography).

2.2.3 Spatial Analysis

Due to its wide range of capabilities and unclear definition, *spatial analysis* (SA) is sometimes referred to as a certain set of techniques or a particular way of processing spatial data since it involves several techniques from different disciplines. For example, in most GIS software packages, spatial data manipulation (e.g., buffering and overlay operation) is often referred to as SA; some refer to SA as descriptive statistics or exploratory data analysis of spatial data, some refer to SA as spatial statistical analysis using advanced statistics, and some refer to SA as spatial modeling [18]. However, here we define SA in a broad sense and from various GIS application perspectives.

SA is a set of techniques or methods that can be applied to raw spatial data and output more useful information to answer questions or explain processes about a real-world phenomenon. Although SA can be applied to any "spatial data" in any space, in GIS the focus is only on applying SA to geospatial data.

In addition to the geospatial data, the non-spatial attributes associated with objects or positions can be considered in the analysis. The techniques or methods of SA remain unchanged when using a different set of spatial data, but the results they produce change according to the changes of objects' locations.

SA is considered as the core of GIS that facilitates GIS users for finding solutions to space-related problems ranging from simple geospatial queries (e.g., finding nearby restaurants) to analysis of a complex natural phenomenon (e.g., finding potential areas of wildfires). As a primary tool in GIS, SA provides features not only for discovering the characteristics of each individual geospatial data set, but also for explaining relationships and interactions (e.g., cause and effect) among multiple data sets within the same study area. As a result, GIS, which can perform SA, is different from CAD, which mainly visualizes spatial data.

Anselin and Getis [19] present a high-level framework for SA commonly found in GIS software packages by categorizing SA's features into four basic functions: spatial data selection, spatial data manipulation, exploratory spatial data analysis, and confirmatory spatial data analysis.

2.2.3.1 Spatial Data Selection

The spatial data selection function enables users to retrieve relevant spatial datasets or observational units from the data storage. The user may search spatial data using specific criteria, dedicated query languages (e.g., SQL), or selection tools provided by interactive user interfaces of GIS software packages. This function makes no changes to the database and no new spatial data generated to the system. Only the requested data are returned to the users. Queries are considered among the most basic operations in SA. They allow users to interrogate the system about any aspect of the geographic data, either by its geometry or by attributes, and to answer simple questions.

2.2.3.2 Spatial Data Manipulation

The spatial data manipulation function contains a number of methods, techniques, and algorithms that convert the selected spatial data to a new insightful data set. Basic data manipulation functions can be divided into five categories: proximity, buffering, overlay, network analysis, and spatial interpolation. Table 2.1 provides

Table 2.1   The Categories of Data Manipulation Function and Samples of Methods

Types	Description	Samples of Operation/ Methods/Algorithms
Proximity	Analyzes locations of nearest entities by measuring distance each pair of them	Distance measurements, nearest neighbor, Voronoi diagrams
Buffering	Creation of a zone of interest around an entity or a set of entities	Buffer, bidirectional buffers, setbacks, causative buffer, measurable buffer
Overlay	Overlays two or more different thematic map layers of the same area to form a new layer	*Vector data*: point-in-polygon, line-in-polygon, polygon-on-polygon, and overlay using spatial operations, such as set-oriented ($\in, \notin, \subseteq, \subseteq$), topological (e.g., meets, overlaps, is inside, covers disjoint, touch) operations *Rater data*: overlay using arithmetic ($+, -, \times, \div$, *exponential, sin,cos.* tan $\square$), logical ($=, >, <, \geq, \leq, <>$), Boolean (AND, OR, NOR, NOT), operations
Network Analysis	Identifies the optimal or efficient routes of a network model	Least-cost path, connectivity, resource allocation
Spatial Interpolation	Estimates the unknown attributes of observed locations from known attributes of multiple locations around the observed location. Mostly, makes sense with field data	Thiessen polygons, triangular irregular networks (TIN), inverse distance weighting(IDW), kringing, density estimation

brief descriptions and examples of each method. For further explanation of these methods refer to [7–10, 14, 20].

2.2.3.3 Exploratory Spatial Data Analysis

The exploratory spatial data analysis (ESDA) function provides insights into the characteristics of data sets. It aids in the identification and description of spatial patterns, reveals the characteristics and statistic parameters of data sets, and helps determine the extent of data dependency and heterogeneity. ESDA, considered as exploratory data analysis (EDA) if applied to non-spatial data, is commonly used as basic statistical data analysis in GIS. Since both ESDA and EDA make no assumptions about the population of the sample data, they are sometimes referred to as "*data driven analysis*" [19]. ESDA techniques include descriptive statistics of EDA (e.g., mean, max, min, standard deviation), spatial related descriptive statistics (e.g., smoothing, trend, spatial autocorrelation, detecting spatial outlier), and statistical graphics (e.g., boxplot, scatterplot, Chernoff faces). An example of a free GIS software package that incorporates ESDA techniques as core operations is GeoDA [21].

2.2.3.4 Confirmation Spatial Data Analysis

The confirmatory spatial data analysis (CSDA) function is for evaluating the evidence, such as theoretical notions or models, gained from manipulating data sets or ESDA. It is also called *model-driven analysis*. CSDA is the most complex SA function, rarely found in commercial GIS software packages because evaluating CSDA results requires specific knowledge within a particular domain. The techniques of CSDA include traditional techniques for hypothesis testing, model fitting, estimation of spatial process models, simulation, and prediction.

2.2.4 GIS Applications

In the early days of GIS, applications were limited within the groups of GIS experts, typically in the areas of land-use and natural resource management. However, today, the use of GIS is widespread to thousands or more applications across diverse domains. This is mainly because most real-world problems are associated with geospatial data, and the modern GIS software packages allow the user to create, store, manipulate, and analyze geospatial data more conveniently than in the past. GIS has become one of the top priority choices for solving geospatial-related problems. Although GIS applications, which range from simple to complex, are too many to be named, they all perform five basic operations: mapping, measurement, monitoring, modeling, and management [9]. Table 2.2 gives examples of major practical applications with six different areas of interests.

Table 2.2   Examples of GIS Practical Applications in Six Areas of Interest

Area of Interest	Examples of Practical Applications
1. Street network	Address matching Vehicle navigation Location analysis and site selection Development of evacuation plan
2. Facilities management	Locating underground pipes or cables Load analysis in power electric network Facility maintenance planning Energy use tracking
3. Land parcel	Zoning, subdivision plan review Land acquisition ownership of maintenance Land-use planning
4. Natural Resource	Forest management Wildlife habitats, migration routes management Wild & scenic river preservation Recreation resources planning Floodplain and land sliding management Wetland preservation Agricultural lands management Ground water modeling and contamination tracking Environment impact analysis Coastal resource management
5. Social	Crime management Health and epidemiology analysis Infrastructure planning Public policy
6. Business	Market and retail analysis Product advertisement planning Strategic and tactical planning Service dispatching planning Customer demographics analysis

2.3 State-of-the-Art GIS

GISs are now widely accepted as powerful tools for storing, manipulating, visualizing, and analyzing spatial data. Many applications can benefit from GIS technology, such as resource management, urban planning, and marketing. Considering current trends in computer technology, GIS is continually growing. In this section, the evolution of GIS, distributed GIS, and Geo Web Services are described, followed by Mobile GIS, three-dimensional GIS, and GIS Interoperability.

2.3.1 GIS Evolution

GIS was born in the late 1960s and has made significant progress since then. Its evolution began from the development of computer mapping in the early 1970s, where its main advantage was in its capability to efficiently update and redraw digital maps. During the 1980s, spatial database management systems were developed by combining computer mapping capabilities with database management capabilities. In that era, the demand for mapped data increased and several world-leading companies in GIS software were established, such as ESRI.

As GIS continued its evolution, spatial statistics and spatial analysis were introduced in the 1990s. The mathematical capabilities integrated with advanced map processing are now available in several modern GIS software packages, which enhance the abilities of spatial decision support systems. Further details on GIS evolution can be found in [22].

Not only has GIS dramatically improved its functionality, its architecture has also evolved. With the increasing availability of ubiquitous computing devices, GISs evolved from desktop GIS to Internet GIS and Mobile GIS and are referred to as "distributed GIServices" [23]. Distributed GIServices increase the number of users by allowing them to access GIS data and manipulate GIS analysis tools interactively over the wired Internet and wireless telecommunication networks without GIS software installation. Moreover, distributed technology can interact with heterogeneous systems and platforms without the constraints of computer hardware and operating systems [24]. Distributed GIServices expand the usage of geographic information into a wide variety of online geospatial applications and services, such as digital earth [25], online mapping [26], water quality monitoring [27], and landslide monitoring [28].

Today, GIS is on everyone's desktop, as well as mobile devices such as PDAs and cell phones. This, in effect, has increased the size of the GIS community, which includes GIS specialists and general users [29]. The current trend for developing GIS is more on distributed GIS, rather than on desktop GIS because of the advancement of wireless technology and standards of the Internet. Due to the high price of software licensing, software training, and software maintenance, GIS vendors break up their proprietary GIS functions into many interoperable functional components and allow users to choose and assemble many geo-processing components over a network based on their needs [30]. This benefits developers to build a customized system at lower cost and generates profits for vendors or service providers because of the massive access to and purchases of services.

2.3.2 Distributed GIS (DGIS)

Peng and Tsou [23] define DGIS as "geographic information services provided through the Internet and allow people to access geographic information, spatial analytical tools and GIS-based Web services without owing a GIS and data." The main purpose of DGIS is to share geographical data and GIS processing tools among developers and users. The evolution of DGIS can be explained in three stages: emerging, evolving, and advanced [31]. In the emerging stage, DGIS provides the lowest functionality and interactivity by allowing users who have access to the Internet to view static maps as graphic images on an HTML document. In the evolving stage, DGIS increases user interactivity by allowing users to query and retrieve customized data, metadata, and other geospatial information, such as all-hazard geospatial data distribution systems [32] and land information services [33]. In the advanced stage, DGIS provides the highest GIS functionality and user interactivity to deliver customized GIS services. Examples of advanced DGIS applications are decision support systems for floodplain, watershed management [34], and environmental monitoring and resource management [35].

2.3.2.1 Basic Components and Functions of DGIS

According to the ISO TC 211 document on Geographic Information, a component is "a physical, replaceable part of a system that packages implementation and conforms to and provides the realization of a set of interfaces" [36]. Distributed components should be plug-and-play, interoperable, portable and reusable, self-describing and self-managing, and able to be freely combined for use [37,38]. There are two important advantages of distributed components. One is the independence from different hardware, network environments, vendors, and applications. Another advantage is the independence from different software environments, database servers, and computer platforms [23]. Therefore, distributed GISs are able to operate on heterogeneous computer platforms and manipulate different types of database servers. Open GIS Consortium (OGC) considers four major components in developing distributed GIS [39]. First, *Viewers and Editors* are the presentation of the system that allows users to view and interact with the system, such as pose a query, view geospatial information including raster and vector data, or resize a map. Second, a *Catalog* is a collection of metadata, which is considered to be information about the data itself. A catalog is useful for search operations. Third, *Repositories* are collections of data. Data repositories are important components in a distributed environment for data discovery and management. Last, *Operators* are components that produce outputs based on user's request, such as a data query, a network analysis, or a spatial analysis.

The following basic functions of DGIS are discussed based on the evolutionary stages mentioned in the previous section. The major service providers are governmental agencies, academic institutions and research centers, and GIS companies [31]. For instance, GeoServNet [40], developed by York University GeoICT Lab, is a Web-based GIS software with unique online 3D visualization functionalities. An earthquake simulation and emergency response was visualized using GeoServNet. In the emerging stage, most functions provide users with an online search and the ability to download static geospatial data. In the evolving stage, most functions also focus on geospatial data and information dissemination with interactive user interfaces and data processing. In other words, users are allowed to make queries on a map including spatial feature selection and feature attribute queries as well as to display, zoom-in/out of spatial information. In the advanced stage, GIS modeling, spatial analysis, and knowledge systems are included in the DGIS functions. In addition, the GIS functionality of this stage is continuously developed in order to deliver services on demand and to solve complex problems.

2.3.2.2 Architecture Models of DGIS

An architecture model provides the framework of the system, including components, functions of each component, relationships, and information flow among components [41]. The architectural model of DGIS depends on the system's purpose. Given that each stage of GIS's evolution has a different purpose, developers should decide what architectural model is suitable for the requirements of the application. Existing architectural models of DGIS are OGC's generic Web mapping architecture [42], three-tier client-server architecture [23], and service-oriented architecture [43]. In this section, we briefly introduce these three architectural models for DGIS. For a more detailed discussion on architectural models, refer to [23,42,43].

■ OGC's generic Web mapping architecture—OGC proposed a general Web mapping architecture that provides a framework to construct an interoperable Web mapping system. The user can search, retrieve, interact, and manipulate geospatial data stored in a distributed environment via Viewers and Editors on the Web browser. This architecture fulfills requirements for the emerging and the evolving stages of DGIS, which aim to provide geographic information through Web browsers. The advantage of this model is that the clientside components are usually platform independent, requiring only an Internet browser to run. This model allows distributed clients to access a centralized server remotely.

■ Three-tier client-server architecture—Three-tier client-server architecture is a simple architecture that is either single-server or multiserver. The single-server architecture has a single map server and a single data server, while the multiserver architecture has multi map servers and multi data servers. The objectives of the multiserver architecture are to handle a large number of user requests and to increase fault tolerance. The first tier, or the "client viewer," is used by users to interact with the system and to view geospatial data. The middle tier includes the application server, which is responsible for making connections between clients and servers. The third tier includes map servers and data servers, which manage all the required data storage. With this architecture, developers are able to create customizable functions for various mapping applications. The advantage of this model is that it allows distributed clients to access a centralized server remotely. However, the clientside components are platform dependent and do not support interoperability.

■ Service-oriented architecture (SOA)—The advent of Web services is leading to a future generation of DGIS development. SOA is "a collection of services with a mechanism to describe, categorize, and discover relevant services, provide described services, and integrate an application based on implemented services" [44]. The idea is that the clients of this architecture can communicate with any application server on the Internet rather than a single application server. In other words, they can take advantages of geospatial data and analysis services from any service provider across the Internet. The advantage of this model is that the clientside components are usually platform independent, requiring only an Internet browser to run. Each system can play both a server role and a client role; therefore it provides flexible access and application on both the client side and the server side.

2.3.3 Geo Web Services

A Web Service is defined by the World Wide Web Consortium (W3C) as a "software system designed to support interoperable machine to machine interaction over a network" [45]. Interest in applying Web service technologies in distributed GIS has tremendously increased in the past several years. Geo Web Services aim to integrate services among heterogeneous geospatial servers and to handle huge volumes of vector data and satellite imagery via both wired and wireless networks. However, it is not easy to provide geospatial services on the Web environment because of the huge volume of geospatial data and the highly complex nature of geospatial data processing. The OGC's implementation specifications, which are Web Map Services Implementation Specification (WMS), Web Coverage Services Implementation Specification (WCS), Geography Markup Language Implementation Specification (GML), and Web Registry Services Implementation Specification (WRS), are to build the Geo Web Services framework. In general, a Web service framework has three components: service provider, service broker, and service requester. A service provider publishes the availability of its services to a service broker using Web Service Description Language (WSDL) documents, and a service requestor performs discovery operations of services from the service broker [46].

There have been many research efforts in Geo Web Services. Research in [47] proposed an architecture of the generic open GIS based on object services and features provided by Common Object Request Broker Architecture (CORBA). Research in [48] implemented a prototype system, GeoServNet, designed for renting software components and providing a service registry for component providers. Research in [30] improved the GeoServNet's capability to offer online 2D/3D analysis and visualization services. Here, we briefly discuss research efforts in Geo Web Service architectures named the MM server, Go-Geo!, and OpenGIS Web services (OWS). For a more detailed discussion on Geo Web service, refer to [49–51].

- MM server—A Web service framework for Geospatial services proposed by [50] called the "MM server" (main memory-based GIS server) supports the vector map, satellite imagery, real-time traffic information, and location information of moving objects. The framework is composed of three essential components similar to a general Web service architecture in which geospatial servers conform to service providers, a geospatial broker conforms to a service broker, and Web-based clients conform to service requestors. The MM server framework provides the integrated geospatial data in the standardized manner by complying with the OGC specifications and referencing the W3C specifications. The result shows that this framework can rapidly serve a huge volume of GML documents on wired network environments, but not on wireless network environments.
- Go-Geo!—Go-Geo! (http://www.gogeo.ac.uk) is an online resource discovery tool that supports geospatial searching by interactive map, grid coordinates and place names, as well as traditional topics or keyword forms of searching [49]. Go-Geo! has been a cooperative effort among EDINA National Data Centre, University of Edinburgh, and the UK Data Archive, University of Essex. The Open Grid Services Architecture (OGSA) is introduced for building standard-based data federation and computation on a grid infrastructure. The benefits from grid infrastructure include supporting resource sharing and high volume data transfer, enabling the online integration of distributed spatial data, and providing data-intensive operations performed on grid resources.
- OpenGIS Web services (OWS)—OpenGIS Web Services (OWS) developed by OGC established a GIService framework based on SOA. OWS architecture illustrates common interfaces, exchange protocols, and services for enterprise-wide interoperability. There are four main components: importer, exporter, registry service, and encoding technique. The importer is a multisource integrated application; for instance, it can be a map viewer, a terrain viewer client, or a discovery client. The exporter is a group of service providers including data services, portrayal services, and processing service providers. The data services serve data, specifically geospatial data, while the portrayal services provide specialized capabilities supporting visualization of geospatial information, such as cartographically portrayed maps, perspective views of terrain, annotated images, and views

of dynamically changing features in space and time. The processing services provide the operations on geospatial data and metadata to provide value-added services. The registry services provide a common mechanism to classify, register, describe, search, maintain, and access information about network resources. The encoding specifications define the way in which geospatial data is transferred in the GIServices workflow. The exporters publish their service in the registry services in which the importers can find the requested services.

2.3.4 Mobile GIS

The advancement of mobile technologies and wireless networks is changing the way people live and work by allowing mobile users to receive digital contents and services anytime and anywhere. The key technologies include GIS, wireless communications, and Global Positioning System (GPS) or other positioning techniques such as cellular network-based positioning. Positioning technology on mobile devices, called mobile GIS, makes the use of geographic data in the field possible. Tsou [52] define the term Mobile GIS as "an integrated software/ hardware framework for the access of geospatial data and location-based services through mobile devices, such as Personal Digital Assistants (PDA) or smart cellular phones via wired or wireless networks." There are two major areas of mobile GIS applications: (1) field-based GIS and (2) location-based services (LBS). The major users of mobile GIS are field workers and consumers of LBS applications [23]. Field-based GIS focuses on field data collection and editing, while LBS concentrates on information services based on the location of the mobile device. In this section, mobile GIS architecture, field-based GIS and its applications, and LBS and its applications are described.

■ Mobile GIS architecture—The general architecture of mobile GIS is composed of the clientside, the serverside, and the communication networks [52–54]. Figure 2.1 shows the high-level system architecture of mobile GIS. The clientside consists of a cellular telephone or PDA equipped with GPS and standard functionalities, such as voice, Short Messaging Service (SMS), and Multimedia Messaging Service (MMS). The mobile device is typically composed of CPU, memory, storage, and input/output interfaces that can display maps and information. The user's current location is determined by GPS or network-based positioning. The serverside consists of middle tier servers and database servers. The middle tier server includes a Web server and an application server. The Web server provides HTTP portal services for Web clients and is responsible for data integrity and the conversion of data into a compatible format. The application server provides the system's functions. The database server is responsible for data storage and management. There are two types of data contained in the database server: spatial and nonspatial data. The communication network consists of the carrier's cellular network

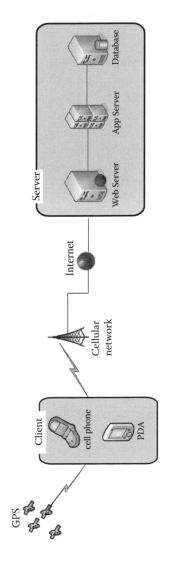

Figure 2.1 The high-level system architecture of mobile GIS.

and the Internet, which assists the exchange of GIS data and services between the client-side and the server-side. For further information about these components, refer to [52–54].

■ Field-Based GIS—The field-based GIS focuses on field data collection, such as collecting soil descriptions in the field. Examples of field-based GIS applications are environmental monitoring and natural resource management, field data collection for ecological/geographic research, and utilities maintenance. Traditionally, the process of collecting field data has been lengthy and not immediately available to the public because it was performed on paper forms. Once data is collected in the field, it is manually entered and validated upon returning to the office, which is often a source of data error and loss [23,52]. Integrating Mobile GIS and GPS allows the data to be represented in the form of a digital map instead of a paper map. The main advantage of employing mobile GIS in the field-based GIS applications is that users are able to add or update new information into the databases [52], which speeds up analysis and decision making by using the up-to-date and accurate spatial data.

■ Location-Based Services (LBSs)—LBSs are "services that integrate a mobile device's location or position with other information so as to provide added value to a user" [55]. Unlike field-based GISs, LBS applications use GIS datasets as the reference maps for navigation or geo-tracking purposes rather than edit GIS data. Enhanced 911 (E911) in the United States is an example of LBS application that enables emergency services to locate geographic location of the caller [56]. Other LBS applications include directory assistance, vehicle and pedestrian navigation, locating nearest landmarks, geo-tracking services, and social interaction services. Generally, there are two different kinds of LBSs—push and pull [56]. Push services automatically deliver information that is activated by an event, while pull services provide information that is requested by users [57]. An example of a pull service is when a user searches for a nearby Chinese restaurant. An example of a push service is when a user receives a weather warning message when weather conditions change. Currently, many service providers increase their business opportunity by creating innovative mobile applications, such as mobile Web, mobile guide and navigation, shopping assistance, and mobile social network. Moreover, the service and content adaptation to users' interests are included for mobile applications to improve the usability of mobile services and satisfy the needs of mobile users [58].

2.3.5 Three-Dimensional GIS

GIS has proved that real-world phenomena can be modeled efficiently using the 2D data models (points, lines, polygons for vector, and grid cells for raster) and their 2D data can be effectively manipulated, analyzed, and represented on a flat map. However, in situations, the 2D data models are inadequate for capturing the details of objects or processes in which the higher-dimensional models are needed, such as 3D models (X, Y, Z) and 4D models (X, Y, Z, and time). Examples of such situations are the applications in the fields of geology, archeology, hydrology, marine biology, and urban modeling. In addition, humans perceive the world as the three-dimensional space. Developing 3D data models is the focus of ongoing research by the GIS community.

3D GIS, representing phenomena using the three-dimensional models, aims to provide five basic functions similar to functions available in 2D GIS: data capture, data structure, data manipulation, data analysis, and data presentation [59]. However, these functions are currently incomplete. The common features usually seen in modern GIS software packages are 3D visualization and animation [60]. ESRI's ArcGIS contains a 3D Analyst module [61] having capabilities to visualize 3D data and analyze surface data. ERDAS provides a GIS module called Imagine VirtualGIS, for displaying 3D environments in real-time [62]. PCI Geomatics provides an add-on module called OrthoEngine 3D Stereo, which offers tools for 3D viewing and feature extraction [63]. These systems have been emphasized only on the 3D visualization aspect but they have yet to offer the complete 3D GIS functionalities.

3D visualization, in fact, is only one component of a practical 3D GIS. The major hindrance causing other 3D GIS functions to be developed slowly is a gap between the representation of 2D and 3D models. The 2D representation models deal with an area of grid cells (in raster) or an area within a boundary (in vector). Its data structure is relatively simple and requires relatively small storage. On the other hand, the 3D representation models must deal with volumes whose data structure is very complex and contains large amounts of data. Both surfaces and detail structures inside an object must be captured in 3D as well as the topological relations with other objects. Once such problems are resolved, 3D GIS will be able to perform volumetric modeling and reasoning, and promote a better understanding of the natural phenomena.

2.3.5.1 Three-Dimensional Data Representation

3D data representation is a key for 3D GIS. Yanbing et al. [64] divided the models into two types: geographical space objects model and geological space objects model. The geographical space objects model mainly deals with the construction above the earth's surface, such as buildings and bridges. The 3D formal data structure (3D FDS) of this category is extended from the 2D formal data structure (2D FDS), consisting of four primitives (node, arc, edge, and faces). Sample models of this type are n-couple model, simplified spatial model (SSM), urban data model (UDM), and object-oriented 3D data model (OO3D Model).

The geological space objects model focuses on the Earth's surface and subsurface objects (e.g., cave, tunnel) as well as natural geological bodies (e.g., stratum, fault).

Example models of this type are triangulated irregular network (TIN) model, grid model, block model, tetrahedral network model (TEN), pyramid model, boundary representation (B-rep) model, constructive solid geometry (CSG) model, and octree model.

2.3.5.2 Three-Dimensional Object Reconstruction

In 2D GIS, the real-world objects can be represented by interpreting information obtained by several data collection techniques, such as surveying and scanning. In contrast, 3D GIS interacts with more information and requires more advanced techniques to reconstruct a 3D representation of each distinct object. Most 3D models available today have been based on CAD software. Although great details of an object can be modeled in CAD, the CAD model still lacks the geospatial and topological properties that are the crucial characteristics of 3D GIS. This prevents 3D GIS from being able to utilize the existing sources of 3D data.

Much research in 3D object reconstruction has been explored, with considerable progress. Several approaches for constructing 3D models, mainly for city models, have been proposed. The quality of the models varies, depending on the quality of data sources, utilized applications, and the required resolution and accuracy. The four general approaches summarized by Stoter and Zlatanova [65] are bottom-up, top-down, detailed reconstructing of all details, and a combination of approaches. Currently, there is no optimal automation approach for 3D construction. Most techniques are semi-automatic or manual, which still require intensive labor. This causes a bottleneck for producing massive 3D objects in 3D GIS.

2.3.5.3 Three-Dimensional Analysis

The analysis function is a core of 3D GIS that clearly separates it from CAD and other modeling software. Unfortunately, the 3D spatial analysis including functions and operations have been slowly developed due to the unsteady foundation of 3D data representations in terms of both geometry and topology. Several 3D data models provide different approaches to capture 3D geometry of objects or phenomena but still lack cohesive agreement on a standard. On the other hand, 3D topology, including adjacency, connectivity, and containment, has yet to be understood [66]. This impedes 3D analytical processes of some applications utilizing 3D GIS, such as applications in the area of geology, archaeology, cadastral, hydrology, transportation, and environmental science.

Two-dimensional spatial analysis cannot simply be applied to 3D spatial data sets for the purpose of 3D spatial analysis. Spatial data (x, y, z) need to be processed simultaneously when any 3D operation is performed. For example, one cannot infer the intersection of two 3D line features by using the 2D intersection operation, since it may return an incorrect result if the two lines do not intersect in 3D space, but do overlap when they are projected on a 2D plane. Examples of 3D spatial operations used in SQL queries of spatial relational DBMSs (e.g., PostGIS, Oracle Spatial), are given in Borrmann et al. [67] and include directional operators (e.g., above, below, north of, south of), topological operators (e.g., touch, contain, equal, inside), metric operators (e.g., distance), and Boolean operators (e.g., union, intersection). The extension of Boolean operators on 3D data, including 3D XOR, 3D union, 3D intersection, and 3D difference, were further investigated in Tet-Khuan et al. [68]. Besides these basic operations for volumetric object manipulation, other basic analytical functions, such as 3D buffering, 3D overlay, 3D shortest route, and 3D inter-visibilities, need to be developed [60] before 3D GIS can be fully utilized.

2.3.6 GIS Interoperability

The term interoperability in the computer or information technology field generally refers to the ability to exchange heterogeneous information and procedures freely among heterogeneous systems or components. According to ISO/IEC 2382–1, interoperability is defined as "the capability to communicate, execute programs, or transfer data among various functional units in a manner that requires the user to have little or no knowledge of the unique characteristics of those units." Interoperability is an important issue for all information systems. This is due to the heterogeneity in all levels of the system structure, which are platform, system, and information heterogeneity [69].

Interoperability greatly benefits geographic information and GIS. It will allow sharing geospatial data and geo-processing tools, integrating among different GIS technologies as well as non-GIS technologies, and encouraging collaboration among different users, developers, organizations, governments, and other units. In the past, conventional GISs had been developed independently without concerns of data sharing and communication between systems. As a result, today we have a variety of GIS software, both open source (e.g., GRASS, MapServer, Geotools, MapWindow GIS) and proprietary (e.g., ArcGIS, Geomedia, MapGuide, MapInfo), GIS software packages with different file formats, geoprocessing operations and tools, spatial databases, and visualization methods. Together with diverse choices of technologies, such as hardware, networking, and programming languages, GIS has a serious heterogeneity problem that prevents realization of its full capabilities by the user. Thus, to have all components of GIS work together and develop conventional GIS, Internet GIS, distributed GIS, Web GIS and mobile GIS, agreements among all stakeholders on standards are required to promote GIS interoperability.

In the United States, the initial standards for GIS interoperability were created through the establishment of the National Spatial Data Infrastructure (NSDI). NSDI was aimed to promote sharing and distributing of geospatial data across all interconnected systems including both federal and private sectors. Federal Geographic Data Committee (FGDC) has driven all activities of NSDI, such as creation of policies, standards, procedures for organizations. The two important standards defined by FGDC that support NSDI are Spatial Data Transfer Standard (SDTS) and Content Standard for Digital Geospatial Metadata (CSDGM). SDTS is a standard for describing spatial data in which it is designed to easily transfer and use on different computer environments. CSDGM is a standard for establishing the names of data elements and compound elements to provide a common set of terminologies and definitions for digital geospatial data.

Another active organization that has been continuously producing worldwide standards for the GIS community is the OGC. OGC is an international collaboration of companies, government agencies, and universities. By using a consensus process, OGC develops specifications to support interoperability for geospatial information, tools, and activities through the Web, wireless and location-based services, and mainstream IT [70]. The most important standards and specifications are OGC Reference Model, Web Map Service (WMS), Web Feature Service (WFS), Web Coverage Service (WCS), Web Processing Service (WPS), Web Catalog Service (CSW), Simple Features (SFS), Geography Markup Language (GML), and Web Service Common (WSC). For updated details of each standard and specification, the reader can refer to[71].

In addition to the two organizations mentioned, a few other organizations—both national and international—involved in developing standards for GIS interoperability are ISO/ TC 211, ANSI (American National Standards Institute), W3C (World Wide Web Consortium), WS-I (Web Services Interoperability Organization), IHO (International Hydrographic Organization), GSDI (Global Spatial Data Infrastructure), CEN (European Committee for Standardization), and DGIW (Digital Geographic Information Working Group).

2.4 Future Directions

Throughout its development, GIS has been evolving in parallel to the advancement of information technologies and continuing to go along with the same speed. Although we have seen slight improvement in the core concepts and theories related to representation and analysis of geographic phenomena, substantial developments on GIS technologies and innovations have drawn attention from geospatial users and industry during the past decade. GIS by itself has sufficient body of knowledge on data representation, models, functions, tools, analysis methods, procedures, and methodologies, but still lacks the ability to optimally utilize the current resources and knowledge due to diversities of data formats, tools, platforms, developers, users, and researchers involved in various domains and applications. In this section, we highlight several current issues and discuss directions of future GIS in the next several years.

Semantic data exchange is a key issue in intelligent and automated systems. Semantic information enhances abilities of systems to interpret and understand meaning of data as well as specific domain knowledge. Ontology is a potential method for representing such meanings of concepts and their relationships. Mutual understanding of the shared concepts and intentions promotes implicit data integration and collaboration among people from different domains. Several fundamental/ top-level (e.g., GIS ontologies and CAD ontologies) and domain-level ontologies (e.g., construction management ontologies) will need to be constructed and agreed on by corresponding communities.

An alternative approach to facilitate spatial data/processes sharing and integration is through the use of Web services. Web-service technologies with the concept of distributed computing have been an important tool for the connected digital world for sharing data, knowledge, and computing resources. Web services wrap heterogeneous spatial information and processes (i.e., GIS operations) under published interfaces defining request and response parameters and make them available to be discovered on the Internet. Using Web services will provide reusability, autonomy, composability, flexibility, and extensibility to all levels of users, from large organizations to individual users, to quickly respond to the changes. Examples of geospatial Web services are MapQuest, Google, Microsoft Virtual Earth, and ArcWeb, which offer data services, such as map services and search services, and basic process services, such as geocoding and routing. Currently, a number of geospatial Web services have sprung up, but they still lack commonalty on request and response parameters. For example, the geocoding services offered by Google require different parameters and data formats from those of MapQuest's geocoding service. Thus, the use of automated chaining or composing ad hoc services based on the concept of SOA requires a defined standard for common request and response parameters to determine quality of services and criteria for judging the outcomes of the candidate services.

Besides agreement on specifications and standards, the next generation of GIS software and tools will be developed based on the concept of ease of use, implementation, and integration. This is partly a result of continuing efforts on GIS interpretability through ontologies and Web services. Unlike the traditional GIS, where geospatial data were generated for a specific task within an organization and GIS operations were designed to fit the needs of groups of professionals, the next generation of GIS will increase availability and accessibility of GIS data, operations, procedures, and software. As such, it will provide opportunities for developers, researchers, and general users to utilize complete and comprehensive resources from a large GIS infrastructure with low cost and less effort but with high quality.

One other aspect is mobile GIS, since mobile devices (e.g., cell phones, smart phones, palmtops, PDAs, and laptops) have become increasingly prevalent. The modern devices are capable of accurately sensing geospatial information, storing a large chunk of data, connecting to the Internet, and performing complex computation. Thus, real-time decision support and location-based services will continue to dominate the geospatial industry by utilizing traditional features to serve the real-time context. Although most existing GIS features can be applied in mobile environments, new data models and techniques are needed to facilitate visualization and computing requirements of the mobile platforms.

References

1. USGS. (2005). Geographics Information Systems [cited 14 March 2008]. Available from http://erg.usgs.gov/isb/pubs/gis_poster/#what.
2. Mark, D. M., Chrisman, N., Frank, A. U., Mchaffie, P. H., and Pickles, J. (1997). The GIS History Project. In *UCGIS Summer Assembly*, Bar Harbor, ME.
3. Centre_for_Advanced_Spatial_Analysis. (2000). Welcome to the GIS Timeline [cited 29 January 2009]. Available from http://www.casa.ucl.ac.uk/gistimeline/.
4. Coppock, J. T. and Rhind, D. W. (1991). The History of GIS. In *Geographical Information Systems : Principles and Applications*, Maguire, D. J., Goodchild, M. F. and Rhind, D. W., Eds. Essex: Longman Scientific & Technical.
5. ESRI. (2008). What Is GIS? [cited 14 March 2008]. Available from http://www.gis.com/whatisgis/index.html.
6. Newell, R. G., and Sancha, T. L. (1990). The Difference between CAD and GIS. *Computer -Aided Design*, 22(3): 131–135.
7. Bernhardsen, T. (1999). *Geographic Information Systems: An Introduction*. Second ed. Danvers, MA: John Wiley & Sons.
8. Jones, C. B. (1997). *Geographical Information Systems and Computer Cartography*. Harlow, UK: Longman.
9. Longley, P. A., Goodchild, M. F., Maguire, D. J., and Rhind, D. W. (2005). *Geographical Information Systems and Science*. 2nd ed. West Sussex: John Wiley & Sons.
10. Burrough, P. A. and Mcdonnell, R. A. (1998). *Principles of Geographical Information Systems*. New York: Oxford University Press.
11. Lo, C. P. and Yeung, A. K. W. (2007). *Concepts and Techniques of Geographic Information Systems* 2nd ed. Upper Saddle River, NJ: Prentice Hall.
12. Hsu, J. Y. (1996). *Computer Networks: Architecture, Protocols, and Software*, Norwood, MA: Artech House.
13. Heywood, I., Cornelius, S., and Carver, S. (1998). *An Introduction to Geographic Information Systems*. White Plains, NY: Longmans.
14. Worboys, M. and Duckham, M. (2004). GIS: *A Computing Perspective*. 2nd ed. Boca Raton, FL: CRC.
15. Yuan, M., Mark, D., Egenhofer, M., and Peuquet, D. (2004). Extensions to Geographic Representations. In *A Research Agenda for Geographic Information Science*, Mcmaster, R. B. and Usery, E., Eds. L. Boca Raton, FL: CRC.
16. Winston, P. H. (1984). *Artificial Intelligence*. 2nd ed. Reading, MA: Addison-Wesley.
17. Ellis, F. (2001). Introduction to GIS: Vector Based GIS [cited 20 March 2008]. Available from http://www.geom.unimelb.edu.au/gisWeb/GISModule/GIST_Vector.htm#fig6.
18. O'Sullivan, D. and Unwin, D. J. (2003). *Geographic Information Analysis*. Hoboken, NJ: Wiley.
19. Anselin, L. and Getis, A. (1992). Spatial Statistical Analysis and Geographic Information Systems. *The Annals of Regional Science*, 26(1): 19–33.
20. Demers, M. N. (2000). *Fundamentals of Geographic Information Systems*. 2nd ed. New York: John Wiley & Sons.
21. Anselin, L. (2006). Geoda: An Introduction to Spatial Data Analysis [cited 20 March 2008]. Available from www.geoda.uiuc.edu/.
22. Berry, J. (2006). Early GIS Technology and Its Expression. *GEO World*, October.
23. Peng, Z.-R. and Tsou, M.-H. (2003). *Internet GIS Distributed Geographic Information Services for the Internet and Wireless Networks*. Hoboken, NJ: John Wiley & Sons, Inc.

24. Montgomery, J. (1997). Distributing Components. *BYTE*, 22(4): 93–98.
25. Goodchild, M. F. (2000). Communicating Geographic Information in a Digital Age. *Annals of the Association of American Geographers*, 90(2): 344–355.
26. Karaak, M. –J. and Brown, A. (2001). *Web Cartography*. London: Taylor & Francis.
27. Jankowski, P., Tsou, M.-H., and Wright, R. D. (2007). Applying Internet Geographic Information System for Water Quality Monitoring. *Geography Compass*, 1(6): 1315–1337.
28. Kandawasvika, A. and Reinhardt, W. (2005). Concept for Interoperable Usage of Multi-Sensors within a Landslide Monitoring Application Scenario. In 8th Conference on Geographic Information Science (AGILE), 26–28 May, at Estoril, Portugal.
29. Berry, J. (2007). A Multifaceted GIS Community. *GEO World*, January.
30. Tao, C. V. (2001). Online GIServices. *Journal of Geospatial Engineering*, 3(2): 135–143.
31. Chen, X. M., Yang, C., and Chen, S. (2005). Evolution and Computing Challenges of Distributed GIS. *Journal of Geographic Information Science*, 11(1): 61–70.
32. Lowe, A. S. (2003). The Federal Emergency Management Agency's Multi-Hazard Flood Map Modernization and the National Map. *Photogrammetric Engineering & Remote Sensing*, 69(10): 1133–1135.
33. Gustafsson, S. (2003). Eulis: European Land Information Service. Co-Operation between Eight National Agencies. *GIM International*, 17(8): 45–47.
34. Dymond, R. L., Regmi, B., Lohani, V. K., and Dietz, R. (2004). Interdisciplinary Web-Enabled Spatial Decision Support System for Watershed Management. *Journal of Resources Planning and Management*, 130(4): 290–300.
35. Tsou, M. H. (2004). Integrating Web-Based GIS and Image Processing Tools for Environmental Monitoring and Natural Resource Management. *Journal of Geographical Systems*, 6(2): 155–174.
36. Iso/Tc211/Wg4. (2000). Cd 19119 Geographic Information-Services (Draft). In ISO/ TC 211-N906.
37. Orfali, R., Harkey, D., and Edwards, J. (1996). *The Essential Distributed Objects Survival Guide*. New York: Wiley.
38. Pountain, D. (1997). The Component Enterprise. *BYTE*, 22(5): 93–98.
39. Open Geospatial Consortium. (2000). OpenGIS Web Map Server Interfaces Implementation Specification, Revision 1.0.0. [cited 28 March 2008]. Available from http://www.opengeospatial.org/standards/wms.
40. Abdalla, R. and Tao, V. (2005). Integrated Distributed GIS Approach for Earthquake Disaster Modeling and Visualization. In *Geo-Information for Disaster Management*, Oosterom, P. V., Zlatanova, S. and Fendel, E. M., Eds.: Springer.
41. Bass, L., Clements, P., and Kazman, R. (1999). *Software Architecture in Practice*: Addison-Wesley.
42. Open Geospatial Consortium. (2000). OpenGIS® Web Map Server Interface Implementation Specification [cited 24 March 2008]. Available from http://www. opengeospatial.org/standards/wms.
43. Cai, G. (2005). Extending Distributed GIS to Support Geo-Collaborative Crisis Management. *Journal of Geographic Information Science*, 11(1): 4–14.
44. Brown, A., Johnson, S., and Kelly, K. (2002). Using Service-Oriented Architecture and Component-Based Development to Build Web Service Application. A Rational Software White Paper.
45. Haas, H. and Brown, A. (2004). Web Services Glossary [cited 24 March 2008]. Available from http://www.w3.org/TR/ws-gloss/.

46. Chinnici, R., Gudgin, M., Moreau, J.-J., and Weerawarana, S. (2007). Web Services Description Language (WSDL) Version 2.0 [cited 24 March 2008]. Available from http://www.w3.org/TR/wsdl20/.
47. Jacobsen, H. A. and Voisard, A. (1998). *CORBA-Based Interoperable Geographic Information Systems*. Humboldt Universitaet Berlin.
48. Tao, C. V. and Yuan, S. (2000). GeoServNet: Renting Geotools over Internet. *Geo-Informatics*, 14(12): 12–15.
49. Boyle, E., Abrahamson, P., Mineter, M., Medyckyj-Scott, D., Reid, J., and Higgins, C. (2005). Go-Geo!—Geo-Data Portal [cited 28 March 2008]. Available from http://www.gogeo.ac.uk/cgi-bin/info.cgi.
50. Kim, M., Kim, M., Lee, E., and Joo, I. (2004). Web Services Framework for Geospatial Services. In *Web and Wireless Geographical Information Systems* (W2GIS). Goyang, Korea.
51. Open Geospatial Consortium. (2003). OGC Web Services Initiative Phase 2 and Demonstration [cited 25 March 2008]. Available from http://www.opengeospatial.org/projects/initiatives/ows-2.
52. Tsou, M.-H. (2004). Integrated Mobile GIS and Wireless Internet Map Servers for Environmental Monitoring and Management. *Cartography and Geographic Information Science*, 31(3): 153–165.
53. Chen, F., Yang, C., Yu, W., Le, X., and Yang, J. (2005). Research on Mobile GIS Based on LBS. In IGARSS'05, Geoscience and Remote Sensing Symposium.
54. Barbeau, S. J., Labrador, M. A., Winters, P. L., Perez, R., and Georggi, N. L. (2006). A General Architecture in Support of Interactive, Multimedia, Location-Based Mobile Applications. *IEEE Communications Magazine*, 156–163.
55. Spiekermann, S. (2004). General Aspects of Location-Based Services. In *Journal of Location-Based Services*, Schiller, J. and Voisard, A., Eds.: Morgan Kaufman.
56. Jiang, B. and Yao, X. (2006). Location-Based Services and GIS in Perspective. *Computers, Environment and Urban Systems*, 30: 712–725.
57. Virrantaus, K., Markkula, J., Garmash, A., and Terziyan, Y. V. (2001). Developing GIS-Supported Location-Based Services. WGIS'2001—First International Workshop on Web Geographical Information Systems, at Kyoto, Japan: 423–432.
58. Goren-Bar, D. (2004). Overcoming Mobile Device Limitations through Adaptive Information Retrieval. *Applied Artificial Intelligence*, 18(6): 513–532.
59. Raper, J. F. and Maguire, D. J. (1992). Design Models and Functionality in GIS. *Computer & Geosciences*, 18(4): 387–394.
60. Zlatanova, S., Rahman, A. A., and Pilouk, M. (2002). 3D GIS: Current Status and Perspectives. Symposium on geospatial theory, processing and applications, 8–12 July, at Ottawa, Canada: 8–12.
61. ESRI. Arcgis 3D Analyst [cited 24 March 2008]. Available from http://www.esri.com/software/arcgis/extensions/3danalyst/index.html.
62. ERDAS. Imagine VirtualGIS [cited 24 March 2008]. Available from http://gi.leica-geosystems.com/LGISub1x39x0.aspx.
63. PCI Geomeatics. (2005). Orthoengine 3D Stereo [cited 24 March 2008]. Available from http://www.pcigeomatics.com/pdfs/3D_Stereo.pdf.
64. Yanbing, W., Lixin, W., Wenzhong, S., and Xiaomeng, L. (2007). On 3D GIS Spatial Modeling. ISPRS Workshop on Updating Geospatial Databases with Imagery & The 5th ISPRS Workshop on DMGISs, 28–29 August, at Xingjiang, China: 237–240.

65. Stoter, J. and Zlatanova, S. (2003). 3D GIS, Where Are We Standing. In Proceedings of the ISPRS WG II/6, IV/1 and IV/2 joint workshop on 'spatial, temporal and multi-dimensional data modelling and analysis', 2–3 October, at Quebec City, Canada.

66. Ellul, C. and Haklay, M. (2006). Requirements for Topology in 3D GIS. Transactions in GIS, 10(2): 157–175.

67. Borrmann, A., Van Treeck, C., and Rank, E. (2006). Towards a 3D Spatial Query Language for Building Information Models. In Proceedings of the 11th International Conference on Computing in Civil and Building Engineering (ICCCBE), 14–16 June, at Montreal, Canada.

68. Tet-Khuan, C., Abdul-Rahman, A., and Zlatanova, S. (2007). 3D Spatial Operations in Geo DBMS Environment for 3D GIS. In Computational Science and Its Applications (ICCSA), 26–29 August, at Kuala Lumpur, Malaysia: 151–163.

69. Sheth, A. (1999). Changing Focus on Interoperability in Information Systems: From System, Syntax, Structure to Semantics. In *Interoperating Geographic Information Systems*, Goodchild, M. F., Egenhofer, M. J., Fegeas, R. G. and Kottman, C. A., Eds. Norwell, MA: Kluwer.

70. Open Geospatial Consortium. (2008). About OGC [cited 25 March 2008]. Available from http://www.opengeospatial.org/ogc.

71. Open Geospatial Consortium. (2008). OpenGIS® Standards and Specifications [cited 24 March 2008]. Available from http://www.opengeospatial.org/standards.

Chapter 3

CAD/GIS Integration
Rationale and Challenges

Omer Akın[*]

Contents

3.1 Introduction

Integrating computer-aided design (CAD, Chapter 1) and geographic information systems (GIS, Chapter 2) is certainly not a new idea, but it is one that has evolved

---

[*] With a contribution from Kristen Kurland.

through time—from a CAD-centered perspective, when both were in an incipient state, to a bilateral perspective, in which both sides have the need for, and seek, data from the other. Yet, whatever integration has been achieved has been through ad hoc, vendor-centric means.

CAD systems place an emphasis on 2D graphics, sketching, and coordinate geometry tools, while GIS focuses on mapping, data management, and geoprocessing. CAD uses mathematical models to generate precise forms, such as circles, arcs, and parallel lines. GIS was developed around an arc/node topology model effective for storing, calculating, and analyzing spatial data.

In the long term, CAD/GIS integration (CGI) is likely to come about through the development of a seamless interface integrating the dissimilar views into that of a single world, a singular perspective. However, there are many different ways that this merger can materialize. In this chapter, we will examine the potential for success for this unlikely merger of the software field's most prominent "odd couple."

3.1.1 Background and Scope of GIS

GIS is a set of computerized tools designed for the storage, retrieval, and analysis of geographically referenced data. GIS uses advanced analytical tools to scientifically explore the spatial relationships, patterns, and processes of cultural, biological, demographic, economic, geographic, and physical phenomena. Because many of GIS's roots are in public sector government agencies, it is often associated with environmental monitoring and modeling of natural features. However, today's users span a wide variety of industries including business (site location, delivery systems, marketing), government (local, state, federal, military), emergency services (fire and police), health (public health, hospitals, health policy and research), economic development, census demographic studies, politics (elections and reappointment), communication, transportation, mining, and urban planning (land use, housing studies, crime analysis).

Conventional database centric applications answer the following questions: who, what, when, why, and how. Although GIS also answers these questions, it also addresses more complex questions, such as "What exists at a particular location?" A location can be described in many ways, for example, using place name, postal or zip code, or geographic references, such as latitude and longitude. GIS examines trends, such as "What has changed over time?" or "What spatial patterns exist?" Unique GIS applications include buffer and proximity analysis, and its strengths are in its spatial modeling tools that combine multiple layers of information.

3.1.2 Levels of Detail and Projections

A complexity of GIS is the level of detail needed and its corresponding projection or coordinate system. Unlike CAD applications that use a rectangular Cartesian coordinate system, GIS uses multiple projections and coordinate systems depending on the scale and purpose of the application. Because GIS features correspond to the surface of the earth and are projected to a flat map, careful consideration needs to go into what projection/coordinate system should be used. Map projection is important, especially when looking at small-scale maps or features of the entire world and producing distortion in one or more spatial properties like shape, area, distance, and direction. Common coordinate systems include the Geographic Coordinate System (North American Datum) used by the U.S. Census, and the State Plane Coordinate system used by local governments. If projections are assigned to GIS layers, the software will adjust them accordingly so that many projections can be viewed together.

GIS studies might include features of the world or a few neighborhood streets. Data sources vary depending on the scale of study. For example, an international organization, such as the World Health Organization, would need data from many countries and its features would include the entire earth. National organizations, such as the U.S. Census, U.S. Geological Survey (USGS), or Environmental Protection Agency (EPA), would need data for that country and its scale might range from states to counties to local census boundaries (e.g., census tracts.) State agencies, such as a state health department, need state level data (e.g., county and municipal boundaries). Local governments, such as a city planning departments, typically need detailed layers, such as buildings, sidewalks, and topography. Because GIS's features are geographically referenced to the world, GIS is seldom used for detailed architectural applications, such as building design and maintenance.

3.1.3 CAD/GIS Integration

Because today's organizations are increasingly complex and often global, they need to facilitate better decision making through multiple graphic and database applications. CAD software allows organizations to view single drawings in great detail while GIS allows them to view multiple sites but with less detail. Both applications serve a specific purpose but organizations often need to visualize or analyze interior and exterior data simultaneously.

There are several important differences that exist between CAD and GIS software that constitute the underpinnings of the need to integrate CAD and GIS:

- CAD is rooted in drafting and has great ability to create detailed geometry.
- GIS is rooted in data management and its strength is in relating geographic features to databases.
- CAD drawings are typically single drawing files (e.g., one floor plan) and well suited for design drawings but are not "database information systems."
- GIS maps often combine multiple feature sets (e.g., streets, buildings, topography) and databases together. GIS also handles many types of data including photos, videos, and sound clips.
- CAD is typically single user, GIS is often multiple users in one dataset (Enterprise Geodatabases).

Real estate and property management is one area that can benefit from CAD/ GIS integration. For example, an organization with many buildings or sites might need to combine detailed floor plan information across multiple buildings for facilities and site infrastructure master planning. Occupancy planning and vacancy analysis requires facility planners to simultaneously view multiple floors in many buildings. Space planners often need to analyze how departments function across many buildings in a campus setting. Real estate managers might ask what city has optimal demographic variables for opening a new office, plant, or retail store. GIS is needed to perform such studies, but detailed CAD drawings are eventually needed during the design phase. An integrated CAD/GIS application is necessary to perform these functions.

Operations management in large, complex organizations often requires access to both external and internal data. A facility manager might ask how utility lines relate to and impact buildings and spaces within the building. Or they might need to know what building or spaces within the building have been checked for environmental compliance or environmental health and safety.

Emergency preparedness and response analysis is another example of the need for CAD/GIS integration. Quick access of data across a campus or city allows emergency response teams to make critical life-safety decisions during a disaster. For example, suppose that a natural gas line has a leak and security personnel need to know which mission critical rooms and equipment are in proximity to the line. CAD software can identify rooms and equipment but does not have the query capability to find features within a distance of the line. GIS is needed to perform such proximity queries.

There is a growing tendency to integrate these in shared applications. This can be explained by the fact that CAD and GIS systems provide information on and deliver representations of the same real-world objects in each phase of the lifecycle. There are several areas of application that illustrate the need for an integrated approach: (1) plan development, (2) visualization, (3) data collection, and (4) location-based services and augmented reality.

ESRI (Environmental Systems Research Institute) is currently the largest GIS software developer. Their *ArcGIS* application, *ArcMap*, reads CAD DGN, .DWG, and .DXF files directly, which is similar to adding a GIS layer in a CAD system. Although CAD drawings cannot be edited directly in GIS, their layers can be turned on or off and they can be exported to create GIS layers, which can be manipulated in GIS. Because GIS's graphic capabilities are often limited to simple and primarily 2D shapes, complex 3D features cannot be drawn directly in GIS and AutoCAD imports are often necessary.

ESRI's GIS *Shapefiles* can be exported as .DXF drawings and imported into CAD. ESRI also provides access to *ArcGIS Server* map services which allow CAD users to visualize and query GIS data without needing conversions in *AutoCAD*. CAD users can work directly with *ArcGIS* map services to add full GIS context within *AutoCAD* sessions. Also some CAD applications, such as *AutoCAD* have integrated GIS geospatial capabilities. Their *AutoCAD Map 3D* and *Map Guide* products provide an early vision for CAD and GIS integration (CGI).

3.2 A Historical View of CAD, GIS, and CGI

The quest for the integration of CAD and GIS began in earnest during the last five years. But like most other movements, the emergence into the limelight came about after decades of experimentation, exploration, and persuasion.

This was upon the backdrop of the reluctance of professional entities and their practices to readily accept change, unless of course a set of very special conditions that surround these movements are met. These conditions include (1) specific traits of individuals who instigate the change, (2) the merits of the movement, and (3) the context within which these changes are to take place (Gladwell [1]). Since the topic of this chapter is not aimed at how such movements are "tipped over," it should suffice to say that, in this case, we are still awaiting the "tipping point." However, as we will argue in the conclusions section of this chapter, this is an inevitable outcome. The only debate is not when but how it will happen.

To lay out the alternative paths to CGI clearly, let's take a quick look at what happened in the preceding decade, as well as the last five years of frenzied activity around the idea of integrating CAD and GIS.

3.2.1 Emergence of GIS

By the beginning of the 1990s, CAD had been around for nearly three decades and gaining momentum in the AEC industry (Chapter 1). GIS as well had a similar emergence into the regional planning and geography fields, starting as early as the early 1960s. This is no coincidence. While cartography is truly an ancient science, digital GIS did not happen until computers became widely accessible and usable. This is the main reason why CAD and GIS emerged around the same time.

Naturally, there were parallel efforts to digitally map the planet Earth. The Department of Defense led as early as the first satellite images, such as these efforts ERTS (Earth Resources Technology Satellite), were gathered in the 1960s. ERTS. Provides digital raster elevation data and related products, which pioneered the documentation of the earth from space. However, organizing all of this information into a usable form took decades to accomplish.

In the Canadian firm of Spartan Air Services, in 1960, some pioneers of cartography and surveying began experimenting with computer technology as an aid to managing spatial data. In particular, Roger Tomlinson, who was asked to produce a map for site analysis for an African country, soon recognized the productivity gains that could be had for such a large-scale application being digitally done. This led to a coalition between Spartan and IBM to develop a bridge between digital formats and geographic data.

After 1962, this pioneering work led Tomlinson to form a liaison with Lee Pratt, head of the Canada Land Inventory (CLI), and to create maps covering Canada's land-related economic sectors including agriculture, forestry, wildlife, and recreation. Pratt too had realized the value of using digital technology to undertake such a daunting task.

The global emergence of today's GIS technology is due, however, to another enterprise under the umbrella of a company called ESRI, this time, in the U.S. Formed in 1969, ESRI has grown into the market giant for GIS that exceeds even Auto's dominance in the field of CAD. *ArcGIS* desktop, which initially emerged from the double motivation of competition and collaboration with *AutoCAD*, is a tool shared by all major CAD vendors. Today, all main cities of the U.S., most governments of the globe, more than three-quarters of Fortune 500 companies, and roughly 10,000 colleges and universities use ESRI products for mapping and spatial analysis.

3.2.2 Emergence of CAD/GIS Integration Efforts

During the early years, CAD and GIS went their own ways for a few decades. This was the incubation period for both technologies, and naturally, their merger could not be even considered prior to their inception and maturation. This is the double sword of technological innovation: once they matured and developed their identity through well-understood practices, data formats, software architectures, use styles, and organizational structures, it became too difficult to simply integrate them. Thus the history of CAD/GIS integration (CGI) does not consist of building a singular idea that addresses both domains, but creating a third entity, which at times seems like a patch, at other times a bridge, and yet at other times like ships sailing past each other in the night. Whether an adaptation or a link conjoining the two systems, this entity needs to negotiate a clear hookup that can integrate data formats, working styles, and even worldviews, seamlessly joining disparate styles.

In the 1990s, when the first papers on the integration issue began to surface (Ervin [2]; McGee [3]), there were already several significant problems recognized due to the dissimilarities that existed between GIS and CAD.

- CAD was specialized to provide functionalities for AEC design and working drawings (double-precision accuracy and the ability to express complex curves and 3D geometry), while GIS did the same for maps for planning purposes (topological data models).
- GIS was based on raster data, incorporating vector data later, while CAD only on vector.
- Due to the divergent practices in each, a gap existed between planning and design.
- Due to data access and conversion, there were productivity and cost-effectiveness problems.

■ When manipulating large map databases, CAD systems quickly reached their performance limit and did not provide the great variety of projections so prevalent in GIS.
■ Early data translators accomplished graphics translation from CAD to GIS, but dropped non-graphic attribute data causing data loss.
■ Early translators accomplished some graphics mapping, yet dropped graphic data from CAD to GIS and most non-graphic data from GIS to CAD; resulting in two-way data loss.

There were also a number of advances realized during this period. A leading innovator was Intergraph, which incorporated *Microstation*,* its CAD software, with *MGE Modeler*, its GIS software. To foster third party software development and combat *AutoCAD* dominance of the field, Intergraph also increased its software's openness by publishing its CAD file format. *Microstation* imported and exported the .DWG data format of *AutoCAD*, enhancing its own GIS application's ability to accommodate data from different sources. Raster-vector mapping was accomplished through built-in functionalities, and attribute data stored in external databases were linked with entities in the CAD or GIS system. Intergraph also had the foresight to support object-oriented (OO) software engineering features.

One of its products targeting CGI was *Plus III's Terramodel*, a 3D land-design tool which supported a suite of applications. *Plus III's Terragration* connected *Terramodel*'s design functionalities with *ARC/INFO* data. *Terramodel* was point-based, enabling accessibility in much richer detail than vector-based CAD systems. With *Terramodel* and *Terragration,* engineers were able to work with large data sets, coordinate geometry, site design, roadway design, sewer and storm water design, hydrology, surface modeling, and volumetric analysis; and perform tasks like coordinating geometric attributes, contouring, profiles, cross sections, 3D views, hydrology, roadway design, photogrammery, and GPS data capture.

ADE (AutoCAD Data Extension) allowed users to bypass the translation bottlenecks and provided more efficient management and access of large GIS files, when these files grew in complexity. *ADE* improved productivity by allowing users to access, modify, analyze, and save only portions of drawings they needed. *ADE* also allowed the change of X-Y coordinates while mapping projections by providing a large number of cartographic coordinate systems and projection conversion facilities.

During this period, one of the first applications that addressed the bridge between surveying and modeling was a 3D land-design tool developed by the

---

* In 1994 Autodesk reported revenue of $419 million while Bentley took in $28 million. In 1997, the year Autodesk shipped Release 14, Autodesk revenue had grown by about 20% to $510 million. Bentley's revenue in 1997 was reported at $115 million, up almost 400% in the same three-year period. During that period of time Bentley went from being approximately the thirteenth largest AEC CAD vendor to being third, behind only Autodesk and Intergraph. In 2000, Bentley Systems purchased the civil engineering and plot management assets of Intergraph and become the second in most AEC segments.

California based Trimble corporation, a surveying firm. Also, Genasys, in Fort Collins, Colorado, developed *GenaCivil,* which ran stand alone or with the company's *GenaMap GIS* application. An example application of *GenaCivil* was the Federal Emergency Management Agency's digital conversion of its flood insurance rate maps with greater speed and accuracy.

The progress made in the 1990s led to import and export of data between CAD and GIS software including general computational capabilities that combine

■ GIS data layers and CAD software,
■ raster data with vector data (triangulated land models with texture maps),
■ geometric data (3D massing) with attribute data (zoning and land use information), and
■ building site and regional scale.

These capabilities brought with them other problems that were taken up, in turn, but not resolved for good. These included converting data conflicts into consistent data, 2D into 3D information, and spectral readings into colors and textures.

All of these advances in computer hardware and software merely began the long process of truly solving the CGI problem; yet they eliminated the early, insurmountable distinctions between GIS and CAD, and affirmed the vision of CGI to be not only necessary, but also feasible. *ADE* and *ArcCAD,* due to their .DWG format compatibilities, provided the opportunity for sophisticated applications, such as OO applications of *AutoCAD-13,* making tighter links between design entities and attribute data. Intergraph users were given the opportunity to achieve precise GIS descriptions in the CAD world of *Microstation. Terramodel* looked forward to being linked with *ArcCAD* and *ArcView* on multiple operating system platforms; and *GenaCivil,* the civil engineering interface of Genasys, allowed those dealing with contextual information to automatically upload GIS data.

> The combination of complex, multi-dimensional digital models with fast computers, high capacity storage devices, telecommunications, and such devices as GPS satellites, multispectral scanners, and immersive display systems offers landscape architects a host of new ways to model and explore landscapes. From all indications, those looking to harness the productivity enhancements of CAD and GIS integration have a lot to look forward to. (Ervin [2])

3.2.3 CAD/GIS Integration

The new millennium ushered in the era of CGI in earnest. This does not mean that there were fewer difficulties in using CAD in the GIS context or vice versa. If anything, greater problems emerged because of the increased participation in the CGI realm and the raising of the bar both by those discovering new areas of application and the greater accessibility to better hardware and software. By the same token, many agencies and companies rose to the occasion, primarily ESRI, the parent company for GIS in the U.S.

In an influential white paper, ESRI [4] argued that significant problems were prevalent when using CAD: (1) to support advanced cartographic requirements; (2) as a data entry tool for a broader user community within the enterprise; (3) in multiuser editing environments; and (4) for sophisticated spatial data handling, modeling, and analysis. Whereas in GIS users are viewed as

- database-oriented in a single seamless database, since GIS manages data over a wide geographic area and offers tools for map projections and handling large data volumes;
- data centric so that they can manage annotations, complex data searches and diverse forms of representing them;
- employing the concept of layering, where each layer can have different data requirements and behaviors or business rules;
- exploiting the relational data organization of GIS, which provides a complete representation of real-world features with the geometric data outside of the primary key and enables flexibility in data query and management.

ESRI summed up the differences between CAD and GIS data models in two key points. CAD is drawing-based and manages data as drawing files or a set of drawing files, and the main product of a CAD system is a paper map; and CAD layers are an entity property, such as color or line type, which is well suited for design drawings but it is not an information system. These descriptions also hint at the competitive flavor with which the mission of CGI was viewed and pursued by one of its principal participants. Similarly, CAD vendors viewed CGI as an add-on, embellishment, or at best, an API of their main software platform. This was one of the central reasons why the integration eventually materialized in an ad hoc and self-centric fashion.

At this time, we see ESRI as one of the major players in developing computer support for CGI. Their 2002 approach created data mapping and translation functionalities to enable the users of GIS to be able to take advantage of CAD functionalities, even if this meant time-sharing between the two platforms or batch processing. They envisioned scalable software covering a breadth of geographic scope and bandwidth. Enterprise information technology was clearly one of the targets among many business applications. Large non-visual data files of corporations dealing with facilities and land holdings were potential clients of CGI.

In order to attract these users to the new applications or revamped old ones, the new tools needed to be user friendly especially for those who were not computer savvy. Yet the range of computation, including complex data structures, search engines, input and output devices, networking potential, adaptability, and visualization methods, had to be accomplished in a domain of bifurcated methods and programs. In other words, the rigors of enterprise information technology applied in the new domain as well.

In fact, ESRI was interested in developing interoperability with all major CAD vendors, including Autodesk's *AutoCAD* and Bentley's *Microstation* through document sharing. The *ProjectWise* environment of Bentley became the focus for file sharing with GIS. In *AutoCAD,* the *CADClient* included as part of *ArcSDE* (Spatial Database Engine) allowed users of *AutoCAD*, and *Microstation* for that matter, to edit features stored in a relational database management system. A proof of concept application with *Graphisoft* focused on sharing objects between the two systems. Standard APIs developed for the application with Bentley were instrumental in bringing *Graphisoft* and GIS together.

Not content with the these advances, ESRI decided to launch a new initiative to significantly enhance CAD/GIS integration. (1) The purpose of this initiative was to: create an intelligent CAD translator, (2) support the latest CAD formats, (3) enhance CAD-like editing in *ArcGIS*, (4) create engineering applications, and (5) develop 3D objects/models. While there is no denying the importance of GIS and CAD individually, they are very different technologies and their integration is no simple matter. This motivated ESRI into developing a long-term goal to improve data and tool interoperability. The *ArcGIS 9* release was launched to address this strategy.

Despite the divergence in data models and operational culture of CAD vs. GIS, data mapping applications like *ArcGIS* specified protocols that fused CAD layers into GIS data models, readjusted these layers to make them compatible with GIS data formats, integrated simple geometric data into these layers, checked their compatibility against GIS business rules, and conducted error checking to clean errant formats that fell through the cracks.

While all of this went a long way toward creating a bridge from GIS to CAD it merely brought to light the inevitable problems of such an approach: the inconvenience of batch processing, data loss in conversion, and the dominance of the host environment in terms of usability and client views. However, one thing was for certain: the integration effort had attained sufficient momentum to become unstoppable.

3.2.4 Emerging CGI Problems

Merely the potential of CGI, aside from those problems that instigated this movement in the first place, set the spotlight on many types of emerging problems. Surveying and engineering projects that deal with large facilities and infrastructure pieces like roadways and bridges; maintaining and operating large facility holdings for corporations, non-governmental organizations, government and private institutions; Defense Department operations ranging from the Coast Guard to air traffic control; navigation and routing plans for first response or emergency situations; highway systems or hard-terrain exploration vehicle design; planning and maintenance of urban settings for healthcare, utilities, and other social programs; and site and network analysis for planning and design tasks.

Take the Coast Guard for example, which has in excess of 40,000 full-time employees, owns about 30 million sq. ft. of building space, operates more than 8,000 buildings at 650-some locations, and controls around 65,000 acres of land. The Charleston Regional Strategic Plan of the Coast Guard addressed the following planning steps for their facilities:

> *As-builts*—verification and compilation of 2D drawings and other survey information
> *Build Model*—3D object models with basic attribute information
> *Model Enhancement*—attribute information compiled by facility assessment
> *Model Utilization*—support strategic planning and capital asset management needs
> *Model Integration*—a database to support (non-)facility management decision making

An immediate step in this process was to create a new database that could accommodate the terrain information as well as the roof forms and other structural features of importance, simultaneously. *ArchiCAD* and *ArcGIS* were used to undertake these tasks. *ArchiCAD* was chosen due to its interoperability features viz *IFC* (Industry Foundation Classes) versions 2.0 and 2.x, developed by the International Alliance for Interoperability (IAI) and Building Lifecycle Interoperable Software (BLIS) project. While helping to improve the Coast Guard's productivity in managing these facilities, the application was living proof of the potential benefits of CGI. Engineers, operations personnel, and planners were able to work together more efficiently. In turn, the AEC documents were embedded in contextual and spatial references where geographic data was enhanced through facility detail information.

Object-oriented 3D models of facilities were used to leverage sound decision making based on resources represented in the knowledge leading to risk management and lower total cost of ownership. The strength of this application stemmed from the schemata developed to portray the facility data in the geospatial context.

Similarly, Autodesk's *Map 3D* software was used to load data into *Oracle Spatial* in order to bridge *ArcGIS* and *AutoCAD* applications onto the same platform. In 2003, just such an application for the 45th Space Wing of the Patrick AFB for installation and real property management used the *MapGuide* of Autodesk to give access to a common installation picture.

This picture was used to "visualize evacuation routes and direct security responses using easy-to-read digital maps … to calculate bomb impacts based on weight and delivery vehicles, and … quickly find optimal locations for command posts during incidents using the directional cordon tool." *ArcGIS* could take a terrain object and put it into *ArchiCAD*; and in turn *ArchiCAD* could bring objects into *ArcGIS*, so they could be viewed in the 3D data environment. "With only a Web browser, authorized users can build maps that incorporate everything from aerial photographs to alphanumeric property data. Plus, over 6,500 CAD and 5,000 as-built drawings are instantly available through the easy-to-use interface. Personnel no longer waste valuable time visiting the GIS office to request maps or CAD drawings, eliminating 75 percent of all custom map requests" (Smith [5–6]).

Utilities represent another potentially widespread, serious CGI application area. Main water pipes residing in one database (GIS) and vulnerable building assets in another (CAD) can pose serious facility maintenance (Lee [7]) and first response problems (Leite [8]). This is one of the arguments for systems like *MapGuide*. Online Spatial Information Systems (OSIS), developed in Calgary, is using a hybrid system consisting of a stitching of GIS and CAD through *MapGuide*, *ARCHIBUS/FM*, and *Cold Fusion*. This compound software environment could map geo-referenced CAD files using both text and graphics. The range of task functionalities included sharing drawings; data cleanup and analysis; and accessing external files, databases, and drawings.

Project Workspace user interface utility is yet another enhancement using a Windows style organization of information that attempts to solve the shortage of screen real estate problem which plagues applications of this kind with multiple "view" requirements. Further problems with such systems are access privileges, security issues, and diverse task functionalities and views. A myriad of less important usability and functionality problems remain.

Enterprise Asset Management (EAM) accounts for a large chunk of the business needs in the AEC sector (around 80%). An EAM system needs to collect and disseminate a body of asset information which can deal with management of (1) *finances* (budgeting, cost tracking, cash flow forecasting, financial reporting); (2) *work* (productivity, scheduling, equipment repair, work duration, activity types, maintenance); (3) *assets* (asset condition, lifecycle cost, performance); and (4) *service* (vendor evaluations, inventory, service). This covers users from the executive levels all the way to the middle managers, field workers, dispatchers, and outsourced personnel. A variety of interface methods are needed as a function of task, context, and user profiles, such as map-based, textual, graphic, virtual, augmented reality, and hybrid systems. These assets are subject to lifecycle management strategies and challenges of technological issues like incompatibility of platforms, complexity, obsolescence, and stove piping.

3.2.5 Emerging CGI Approaches

As the potential solutions for CGI are put into place, problems in addition to ones that were identified above will emerge. Together these will provide the challenges for strategic approaches intended to take on the long-term CGI solution once and for all. There are three potential strategies: (1) *linking*: data transfer and batch processing; (2) *reconfiguring*: specialized APIs or software patches that modify one system (say GIS) so that it could also function like the other (CAD), even if in limited ways; and (3) *rebuilding*: developing new software that incorporates data and functionalities present in both software worlds. The recent history of CGI evolution shows a trajectory that starts with linking, moves on to reconfiguring, and finally ends with building.

Linking strategies respond to urgent needs that arise from the fields where institutions, agencies, and ordinances require tasks that have to merge CAD and GIS data. Facility management, security assessment, urban design, and asset management are some of the areas of application. Bill Miller, design manager with ESRI says "Architects or planners involved in urban regeneration need to look at a 12- to 20-block area, and they need to look at a wide range of factors, such as land use suitability, noise, security analysis, green roof analysis, and economics" (Sipes [9]).

Design of a hydraulic system at the urban or city scale involves both GIS level metadata and CAD level detail data to be considered in the same view. Potentially this can bring about orders of magnitude of profit to the engineering firm, since they can deal with the lifecycle of the design delivery process within a unified design medium.

In the early work, during the 1980s the question posed was how to use CAD software to do GIS-like tasks, from geographic data entry to automated map creation, and vice versa. Due to the work done later, described in Sections 3.2.3 and 3.2.4, attention was refocused on the task of integrating the two technologies. This requires the integration of the underlying task ontologies—entity types, relationships, and operations performed by them—of the two systems into an integrated whole. Only then can there be meaningful exchange between the worlds of the designed entity and its natural context, uninterrupted by different scales, and a seamless exchange between the various stages of the AEC processes. Of course the data model, interface environment, and the touch-and-feel of the system must be combined without explicit seams between GIS-domain tasks and entities versus the CAD ones.

It is not a wonder that such a vision invites many challenges: GIS entities are laid out in small-scale graphics, emphasizing attribute descriptions and indexed access of abstract objects to perform cartography, spatial analysis, map generation, and network analysis. In contrast, CAD systems place a premium on 2D graphics, sketching, geometric modeling, dealing with site design, construction documents, scheduling, and facility operations and management. Professional practices corresponding to each world are also entrenched in and through their respective technologies. In order to realize CGI, the complex organizational-procedural handicaps presented by these two domains need to be overcome.

These two views of the world, while different, are complementary aspects of the physical world that has been decoupled only as a result of professional divisions and compartmentalized development of their tools Merging the existing capabilities, through linking or reconfiguring strategies, only amplifies the domain differences and focuses a lot of energy on patching and filling the gaps in an ad hoc manner. This does not create a smooth and efficient work environment. These systems introduce spurious processes addressing the idiosyncrasies of the patch-job rather than the true nature of the task at hand. As a result they are clumsy and slow in handling the data. Batch processes require large amounts of screen real estate.

Three key integration challenges remain: file translation between vector and raster data in a manner invisible to the user; direct input–output of data and their models into the realm of the combined functionality; and access, retrieval, modification, and maintenance of a shared database.

3.2.6 Emerging CGI Solutions

Naturally, one of the key players in the CAD/GIS integration arena has been Autodesk. *AutoCAD*'s *Map 3D*, *MapGuide* and *Envision* all deliver some level of CGI functionality. *Map 3D* represents contours, surface triangulation, face and slope directionality, slope and elevation, and watershed designations. It is also able to input and output hybrid data, perform precision mapping, and analyze attribute data. *MapGuide* helps publish the data developed in *Map 3D* on the Internet. *Envision 8* (or *OnSite Desktop*) supports tablet technology for reading digital data.

GeoGraphics of Bentley's *Microstation* family of software is another significant integration tool that helps perform spatial analysis, build maps, and encode attributes of entities in data files, tasks favored by Department of Transportation engineers. Finally Intergraph's *GeoMedia* is an example of a tool with the CAD look and feel and direct access to major data formats needed for this integration.

Graphisoft's *ArchiCAD* supports Boolean geometry suitable for organic forms, moving it a step closer to integration with ESRI's *ArcGIS*. ESRI's *ArcGIS* is the dominant product for CGI. Each new version of the software introduces superior CAD editing tools in the GIS environment. It supports directional CAD-GIS translation, improved geo-database features, geo-processing, superior annotation and labeling, realistic visualization (photo textures that are true 3D models), a file conversion utility, a CAD translator, tablet support, and the ability to read files in all major CAD vendor formats—*AutoCAD* (DWG/DXF), *MicroStation* (DGN V8), *ArchiCAD*, and *SketchUp*. Several other CGI solutions are also noteworthy:

Avatech Solutions' *OSIS* (Online Spatial Information Systems), that combines Autodesk's *Map*, *MapGuide*, and *ARCHIBUS/FM* to create a system for facility managers.

CompassTrac from *CompassCom* is for users who need to view CAD and GIS data, but don't need to modify it.

Haestad Methods' *GIS Connect* (http://www.haestad.com) is an add-on program for *AutoCAD* that provides access to the data management and geospatial analysis capabilities of *ArcGIS*.

Any*GIS from Hitachi Software Global Technology is an object-oriented framework that lets you integrate all formats and sources through a common Web interface.

Pitney Bowes' *MapInfo Professional 7.5* enables you to combine data from different sources in the same map window; *MapX Mobile Tool* exports map windows to a Pocket PC.

There are also important developments in the area of public safety. *VisiNet GIS Link*,™ a new CGI product of TriTech, helps integrate spatial data with street address information with ease. Based on ESRI's *ArcGis Engine* 9.1 and Microsoft's .NET technologies, *VisiNet* takes the complexity out of synchronizing street information in a spatial database. When new information is entered, the software automatically updates all affected areas in the database and corrects formatting, redundancy, and data loss problems.

3.3 Rationale and Challenges

As we enter the final stretch in the effort to find a workable integration of CAD and GIS, recognizing its urgency both in the private and government sectors, it can be helpful to review the rationale and challenges surrounding it.

Both CAD and GIS have overlapping and complementary features. Their main commonality is the ability to represent physical entities in space. They overlap through the representation of man-made entities at different levels of abstraction, such as buildings composed of a complex collection of several million solids in CAD versus simple rectangular prisms in GIS. On the other hand, the capabilities of GIS to represent large-scale natural environments and to organize and search attributes of these natural elements in a relational database application complement the geometric representation capabilities of CAD.

The scope of CGI is illustrated by the interrelated domains served by either or both application types (Figure 3.1). CAD systems are applications tailored to design with high precision requirements, while GIS is a response to the natural sciences'

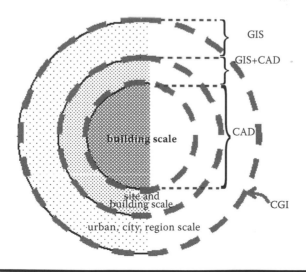

Figure 3.1   Schematic of CAD, GIS, and CGI and the tools that represent them.

need for capturing large space configurations. GIS incorporates topological representations, based on global coordinate systems, for the sake of analysis of natural entities; while CAD systems rely on local Cartesian coordinate information.

Other differences include data format, spatial scale of representations, and semantic issues. These differences become acute when users need to work with information from the other domain. CAD users find it difficult to accurately register a building to its environment while retaining all GIS information. GIS users find it difficult to configure the correct scale and position for imported entities while retaining all CAD information. Underlying this issue is the age-old issue of merging or interconnecting relational and OO databases. Relational database models reach their limit "when extending information models to support new data types, extensible data types, direct support of objects, and deploying in distributed environments with complex operations" (Objectivity, Inc. [10]). OO database models become cumbersome when the number of data relation levels exceeds even a modest number of nestings, and in handling inheritance and subsumption relationships with complex data. Bridging the divide is neither technically nor politically easy. Until the demand for CGI applications become stronger, a widespread solution to the problem does not seem to be in the horizon.

These are the challenges that have become the drivers for the investigation of efficient and seamless integration of these systems. Software that has had an impact in the emergence of a lasting CGI strategy comes under two categories: data translation and APIs.

> *Data translation software.* It has been shown in a diverse set of applications that GIS DEMs (digital elevation models), and DRGs (digital raster graphics) could be brought into *AutoCAD Map* to digitize landscape districts. Guthrie's *Cad2Shape* was used to covert *AutoCAD* DXF/DWG files to *ArcView*/ESRI *Shapefile* format. Data Transformer Extensions of Autodesk are add-ons for *AutoCAD Map* and *Land Development Desktop*. An *ArcView* GIS extension, *Blue Marble*, achieves this through *CADWriter*. Direct read does the file conversion in real time, rather than through file conversion. In all of these data translation tools the problem of data loss and error-prone mapping operations remain.
>
> *API.* ESRI's server extension, *ArcSDE*, provides accessibility to client software through an embedded API. Bentley also introduced an API integrating AEC/GIS with ESRI *ArcGIS*. A similar third party add-on for integration, enabled by Bentley's open source policy (Section 3.2.3), was *GeoGraphics*, which also uses an open *Oracle Spatial* database storage system. A separate API, *ArcScene*, is used to render and query 3D objects, achieving greater flexibility and realism in these models.

Yet, the CGI capabilities represented above fall far short of the integral design environments the two systems, CAD and GIS, represent on their own. In bridging this gap, language like *Java* and *Active X* architectures, for example, can become instrumental, through open GIS compliance, since they are indifferent to GIS or CAD data types. ESRI, Autodesk, and Bentley Systems, are major players in the *ArcGIS* strategy to achieve CGI. ESRI's Bill Miller says:

> Today, the limitation has more to do with the hesitancy of users to change the way they work. Changes are difficult in any situation, and most of us find it hard to master new technology. Many CAD users find the whole idea of thematic mapping and spatial data a little daunting, and many who tried to integrate GIS into their work have given up out of frustration. GIS users who are used to thinking in terms of points and nodes find CAD programs overwhelming because of the demand for detail and precision. Most users need a reason to change before they actually make the leap and learn how to integrate CAD and GIS. There used to be a huge gap between CAD and GIS ... But now it's probably more of a collision zone than a gap.

While a great deal has been accomplished through the myriad of software applications, several important motivations to further innovate remain. There is still a very long list of constituents who need CGI in order to deal with large stocks of buildings that they own, provide for, or manage: planners, real estate developers, engineers, architects, builders, suppliers, utility contractors, permitting agencies, project managers, building managers, maintenance contractors, owners, bankers, brokers, insurance agents, attorneys, accountants, and security and emergency response organizations. Through this very involvement, these constituents add further information to a facility's information base, making the problem of CGI even more challenging. Without an integrated environment, this information is lost or has to be reentered in each new stage of work (Figure 1.4).

Today we are beginning to find better answers to the central issues in CGI. At once, we are seeing more sophisticated and seamless integration of the legacy systems both in CAD and GIS, as well as comprehensive data models (geospatial models and BIM) that will have the capability to breach the conventional barriers between CAD and GIS. One of the ongoing demands in the field, an integrated environment to counteract security and emergency threats, continues to fuel the fire of integration. India and China, with their hyper growth rates, render these innovations in information management a mandate for sustainability of BIM.

International agencies and global awareness of security issues have led to many international examples of CGI. *CORNET,* a system being developed by the Ministry of National Development, Singapore, is providing the lifecycle information needed for buildings and the automatic code compliance checking for the construction industry. Private companies realize the added value from integrated building information models that can revolutionize the building industry through risk analysis, visualization, space planning, requirement programming, egress simulation, owner/tenant relations, construction scheduling, energy simulation, and facility operation and management.

To achieve these ends, coordination of standards organizations is a must. The CAD-GIS Interoperability Working Group of the Open Geospatial Consortium, Inc. (OGC) has been spearheading advances toward the integration of AEC/CAD/GIS (Figure 3.1). OGC members are working on projects, such as:

- Integrating the *OpenGIS Geography Markup Language* (GML) with Building Information Modeling (BIM)
- Helping to develop *TransXML*, a broadly accepted public domain *XML* and *GML* schema for data exchange
- *CityGML*, a unified three-dimensional (3D) city model based on *GML*, and a *Web 3D* specification that integrates CAD, GIS, and the *OpenGIS Web Terrain Service Specification*
- Unified Modeling Language (UML)-based Use Case Articulation for Industry Foundation Classes (IFC)
- Developing interoperability between geospatial and BIM environments: *Sensor Web Enablement*; *GeoProcessing* workflow using *SOAP*; *GeoDecision* support services; multilingual *OGC* Web services; *GeoDigitalRights* management; spatial media (*Mass Market Geospatial* Web services); and *Open Location* services

Also, coordination with other standards organizations is critical, such as the IAI, parent institution to IFC; the Open Consortium for Real Estate Agencies; National Institute of Building Sciences (NBIMS); and FIATECH, an organization dedicated to the integration and automation of the effective execution of capital projects through advanced technology. There is a concerted effort to leverage the Internet and the open source movement to improve the standardization efforts in CGI. In spite of the usual impediments (implementation costs, overcoming inertia to motivate change, new training and documentation needs), the potential benefits offered by the market benefits of CGI far outweigh the costs. Conversely the costs of not achieving interoperability are estimated to be around $15 billion a year [14].

3.4 Summary and Conclusion

We envision CGI, in its mature form, to employ all information relevant in a facility's lifecycle persistently, accurately, and with interoperability. Information belonging to different granularity, scale, and components related spatially or through intra-connectivity need to be operationally combined. Whether it is tracing the path of a main supply line that crosses CAD-GIS boundaries, or correlating the signals of a GPS receiver with those of equipment sensors, there must be shared and standard representations, ontologies, and protocols that eliminate seams for the user.

The barriers between professionals and practitioners in the facility and infrastructure sectors resulting from fundamentally different views provided by CAD and GIS also need to be eliminated. Ontologies used in each domain must be streamlined to facilitate ease of communication and collaboration between professionals.

While a study of landscape and planning professionals shows that CAD and GIS applications dominate usage (Paar [11]), the current practice of manually traversing GIS-CAD boundaries is inefficient. This requires strategies for parameterization, procedural definition, topology, and constraint management within a new semantics that corresponds to CGI applications (Ossterom et al. [12–13]). Ever present digital technologies provide new opportunities through Web and emerging powerful interoperability applications. As illustrated by consortia like Open Geospatial Consortium (OGC, Cote [14]), opportunities for CGI are becoming easier to realize.

While there are many significant similarities between the two realms (geometric representation, data structures) there are also big differences (spatial scale, time scale, accuracy granularity, man-made vs. natural objects). In a CAD document, we find spatial, often 3D, models of components (pipes, walls, windows, etc.), where as these are shown as single lines or nodes, if at all, in GIS. The integration of the two needs the development of new representations that can incorporate different information granularities.

These challenges are not limited to technical innovation. Professionals who function in one domain or the other, not to mention vendors of software, and facility system and component manufacturers, use divergent nomenclature, standards, methods, tests, and legal procedures. If CGI is going to succeed, these communication barriers need to be overcome. A first step would be the development of component, system, and content ontologies that straddle both CAD and GIS. Existing models, such as IFC, and similar AEC applications, such as *aecXML*, *gbXML*, can serve as benchmarks and testbeds for the new models. In turn, CGI ontologies can form the basis for merging existing tools, converting one to cover the functionalities of the other, or inventing all together new ones.

The rationale and needs of CGI have given rise to complex challenges for integration. These span technical, commercial, and social issues, which are treated in the following chapters. Fundamental challenges include integrated geometric data structures and topological support, harmonized semantics of the concepts, and integrated data management (Oosterom [11]).

3.5 Additional Sources for the GIS/CAD Reader

Below we provide sources that can further expand on the points made in this chapter and supply new venues for the reader's individual interests.

▪ Abdul-Rahman, A., Zlatanova, S., and Coors, V. (2006) "Geoinformation and Cartography, Innovations in 3D Geo Information Systems" Lecture Notes 10.1007/978-3-540-36998-1_19.
▪ Autodesk Geospatial Solutions, Geospatial Resource Center http://lp.adskhost. com/index.php/1335-overview.html?CMP=KNC-util-geosp-aw-rc-over-1.

- Bacharach, S. (2006) "CAD/GIS/BIM Integration through Standards." Outreach and Community Adoption, Open Geospatial Consortium, Inc.
- Bacharach, S. (2006) "Integrating CAD and GIS Data to Enable Better Intelligence," Outreach and Community Adoption, for the Open Geospatial Consortium, Inc., 1 Jun.
- Judd, D. D. (2003) "Benefits of GIS Integration with an Enterprise Asset Management System," Ala Carto Consulting, CA, 20 March.
- Kelsey, P. (2007) "Column: Survey/CAD/GIS Integration—It's About Time!" 25 November.
- Kinne, P. (2008) "Integration of CAD and GIS," *Karelcad*, Brisbane, 07 April.
- Letham, G. (2003) "CAD + GIS integration and the A-Z's of AutoCAD Map," 13 December.
- Luccio, M. (2007) "Autodesk Further Integrates CAD and GIS" *GIS Monitor* Reed Business Geo, Inc., Frederick, MD 21702 USA 22 February.
- Maguire, D. J. (2003) "Improving CAD-GIS Interoperability," *ESRI,* Winter 2002/2003.
- Maguire, D.D.J. (2007) "Improving CAD-GIS Interoperability" from http://www.esri.com/news/arcnews/winter0203articles/improving-cad.html.
- McAdams, M. (2006) "Leading Public Safety Software Provider Unveils New GIS Integration Engine, TriTech Software Systems to Provide New Mapping Solution," TriTech Software Systems 858/799-7827, San Diego, CA, 26 May.
- Peachavanish, R., H. A. Karimi, et al. (2006) "An Ontological Engineering Approach for integrating CAD and GIS in Support of Infrastructure Management." *Advanced Engineering Informatics* 20 (1), 71–88.
- Reid, H. (2003) "CAD, GIS Integration: Why Would You Want to Do This Anyway?" *Directions Magazine*, 26 December.
- Turkowski, R. (2008) "New Specifications Seek CAD/GIS Standards," Software Solutions Group, Intel Corp. Issue Date: January 2008, Posted On: 2/1/2008.

References

[1] Gladwell, M. (2000) *Tipping Point: How Little Things Can Make a Big Difference*, Little Brown and Co, Boston.
[2] Ervin, S. M. (1992) "Designing with Maps: Integrating GIS and CAD." http://www.gsd.harvard.edu/-serving/giscad/giscad.html.
[3] McGee, D. (1995) "Developments in CAD and GIS Integration: Bring Design and Mapping Professionals the Best of Both Worlds."
[4] ESRI® (2002) "GIS and CAD—The Right Tool for the Job" white paper, Environmental Systems Research Institute, Redlands, CA. June.
[5] Smith, S. (2003) "CAD/GIS Integration at the U.S. Coast Guard," *GISWeekly Review*, 3 November.

[6] Smith, S. (2003) "Mission Possible: GIS and CAD Data Integration Furthers Space Wing's Mission; Making that CAD/GIS Connection with GISConnect," *GISWeekly Review*, 27 October.

[7] Lee, S. H. and O. Akin, (2008) "Shadowing Tradespersons," *Journal of Automation in Construction*, in print.

[8] Leite, F.B. Akinci, O. Akin, and J. Garrett, (2006) "Supporting the Decision-Making Process of First Responders in Building Emergencies: The Need for a Vulnerability Representation," European Group for Intelligent Computing in Engineering conference proceedings.

[9] Sipes, J.L. (2004) "AEC from the Ground Up: CAD and GIS Integration New Tools Help Combine Drawing and Analysis," *Cadalyst*, 1 July.

[10] Objectivity Inc. (2008) "Object Oriented Database vs. Relational Database," an Objectivity Inc. white paper, http://www.objectivity.com/pages/object-oriented-database-vs-relational-database/default.html.

[11] Paar, P. (2006) "Landscape Visualizations: Applications and Requirements of 3D Visualization Software for Environmental Planning." *Computers, Environment and Urban Systems* 30 (6), 815–839.

[12] Oosterom, P. J. M. v., and J. E. Stoter (2005) "Bridging the Worlds of CAD and GIS" in *Large-Scale 3D Data Integration: Challenges and Opportunities* by D. P. E. S Zlatanova. CRC Press, Taylor & Francis Group, Boca Raton, FL, pp. 9–36.

[13] Oosterom, P.v. (2004) "Bridging the Worlds of CAD and GIS" Px Point on Line Part 1 of a Series on CAD-GIS, Delft University of Technology, June 17, http://www.directionsmag.com/article.php?article_id=601&trv=1.

[14] Cote, P. (2007). OGC Web Services Architecture for CAD GIS and BIM, Open Geospatial Consortium Inc.

Chapter 4

Interoperable Methodologies and Techniques in CAD

Semiha Kiziltas, Fernanda Leite,
Burcu Akinci, and Robert R. Lipman

Contents

4.1 Introduction

Interoperability has been a longstanding problem within the architecture, engineering, construction, and facility management (AEC/FM) industry due to high fragmentation with tight dependency among project participants. Various parties, such as architects, engineers, contractors and suppliers, play a role in generating project data throughout the lifecycle of a construction project. Each party utilizes task-specific software systems, and if interoperability between such systems is not maintained, this situation might result in non-value adding tasks, such as manually reentering data, utilizing duplicate systems and models, and version checking. Lack of or inadequate interoperability results in data and transfer problems and duplication of business transactions across multiple software applications used between architects, owners, engineers, suppliers and facility managers (Young et al. 2007). It has been claimed that such interoperability problems within the AEC/FM industry reached to $15.8 billion in 2002 (NIST 2004), of which approximately $500 million was for manually reentering data. In addition, it has also been identified that for each construction project, the interoperability issues cost about 3.1% of a project's total cost on average (Young et al. 2007).

Many definitions of interoperability exist. It is defined as "a series of data exchanges between computer applications or other software components" by the International Alliance for Interoperability (2009); "exchange and management of electronic information, where individuals and systems are able to identify and access information seamlessly" by the National Institute of Standards and Technology (2004); and as "the ability to manage and communicate electric product and project data among collaborating firms, such as architects, engineers, contractors, owners and building product manufactures" by Young et al. (2007). All these definitions of interoperability suggest that information among applications should be accessed and exchanged without reentering, reformatting, or transforming.

Many data standards and specifications have been developed within the AEC/FM industry to reduce the problem of inadequate interoperability and to streamline exchange of information consistently. The idea behind these data exchange specifications and standards is to define a standard schema for a neutral file or data structure format so that task-specific applications can read data presented in this standard format and generate a similar format of data to be exchanged with other software systems. Having a standard data description and format relieves the software vendors and users from writing specific translators to transfer data among different software systems, and hence streamlines the data exchange process and minimizes interoperability issues. Some of the efforts defined what should be represented and how they should be represented, but did not pass a formal review process from a standards organization (e.g., ANSI, ISO). To make the differentiation between data schemas in terms of standardization, such efforts that have not passed a formal review are called "specifications," and the ones that were standardized by an organization are called "standards" throughout this chapter.

Initial efforts on developing a neutral file started around the late 1970s for exchanging geometry and topology information. Examples of early industry-accepted data formats are Drawing eXchange Format (DXF) initiated by Autodesk, and Initial Graphics Exchange Specification (IGES), which evolved from the U.S. Air Force Integrated Computer Automated Manufacturing Program (ICAM), and was led by large CAD users, such as Boeing and General Electric. These data formats addressed some of the problems associated with interoperability among various engineering and design applications by enabling the exchange of predominantly geometric information (Eastman 1999). However, they also had several limitations, such as inadequate representations that incorporate various functionalities in engineering applications, sometimes subjective mappings of represented entities by data translators, and insufficient conformance testing infrastructure (Bloor and Owen 1995). Standardization efforts continued in the 1980s with STEP (STandard for the Exchange of Product data model), which incorporates object-oriented modeling concepts and enables exchange of computer interpretable product data. Incorporation of object-oriented modeling concepts has played an important role in capturing and exchanging semantic information (such as relationships, properties of products) related to represented products in a digital environment.

Building on STEP core representation models, many data exchange standards and specifications have bloomed, targeting the exchange of data specific to a domain, such as steel and precast concrete (which are also known as "aspect models'), or targeting the support of various domains and phases of a facility. Examples of aspect models are CIMSteel Integration Standards (CIS/2), Automating Equipment Information eXchange (AEX), and Building Information Model for Precast Concrete (BPC).

With the rising need to perform required AEC/FM related tasks in digital environments, studies continued for the development of semantically rich building information modeling standards, where semantics provide meaning to the geometric representations. These standards aim to have more semantics in a given model and enable cross-domain data exchange. One example of such larger cross-domain data exchange standards is Industry Foundation Classes (IFC). There are also standards developed for exchanging domain specific information (other than product data) that are not specific to the AEC/FM industry, but can be utilized within it. Examples of such standards include Sensor Model Language (sensorML) or Land eXtensible Markup Language (landXML).

The main purposes of this chapter are (1) to briefly overview the history of computer-aided design, from 2D drawings to building information modeling (BIM) and semantics; (2) to give an overview of the main data exchange standards developed to support interoperability within the architecture, engineering, construction and facility management (AEC/FM) domain; (3) to evaluate and compare the overviewed data standards; and (4) to discuss the current status of interoperability in the AEC/FM domain and make a projection toward the future.

4.2 History of CAD from Drawings to Building Information Modeling and the Role of Semantics

Computer-aided design (CAD) has been an active research area for decades. Initial efforts are dated to the 1960s, with Sketchpad, developed by the MIT Lincoln Lab (Eastman 1999). Sketchpad was conceived as a drawing assistant for both technical and artistic purposes. The way the program organized its geometric data pioneered the use of "objects" and "instances" in computing (Sutherland 2003). During the 1970s and early 1980s, Charles Eastman, then Professor at the School of Architecture at Carnegie Mellon, was developing a database of several hundred thousand architectural elements which could be assembled and drawn on screen into a complete design concept (Bozdoc 2004). The early work from Eastman's research group at Carnegie Mellon was one of the first parametric modeling efforts, as they developed operations that included spatial transforms, spatial set operations, and Euler operators, which were required for defining new parametric shape primitives (Eastman 1999).

The 1980s can be summarized by advances in parametric modeling (e.g., CATIA and ArchiCAD) as well as a wider distribution and adoption of computer-aided drafting and design technologies by the marketplace (e.g., AutoCAD). In the 1990s, 3D Studio was released (Bozdoc 2004), which is still one of the most widely used off-the-shelf 3D animation programs. Even though animations and renderings created using 3D Studio are photorealistic, modeled objects do not contain "domain" semantics, which is a key concept for the current drivers in building information modeling.

Semantics in building information modeling can be understood as objects with meaning. In other words, an object representing a metal stud wall will know that it is representing a metal stud wall, the dimensions of the wall, what materials the wall is made of, when and where it will be built, what other building elements it is connected to, what other elements, such as windows, that it contains, which two spaces it separates, and so on. Unlike the photorealistic (with no semantics) approach in which the wall would look like the intended wall type, a semantically rich model will contain information about that object, such as its type and specifications. Such semantics are being used in Building Information Models (BIM), a term defined by the National Institute of Building Sciences (NIBS) Facilities Information Council (FIC) as "a computable representation of the physical and functional characteristics of a facility and its related project/lifecycle information using open industry standards to inform decision making for realizing better value" (NIBS 2007). The National Building Information Model committee defines BIM as "a standard repository of information for the facility owner/operator to use and maintain throughout the lifecycle of a facility (NBIMS 2007)." The basic premise of BIM is collaboration by different stakeholders at different phases of the lifecycle of a facility. Although this is a term widely used today, early notions of what is now understood as BIM date back to the 1970s (Eastman 1975).

Several case studies are described by Eastman et al. (2008) in which BIM has played a significant role. These cases, along with many others, represent the pioneering experiences of professionals, such as owners, engineers, architects, contractors, fabricators, and others in the application of BIM in construction projects. Another driver of BIM is the U.S. General Services Administration (GSA), which according to GSA's Office of Government-wide Policy (2006) is the largest lessee of building assets in the United States, with 169 million square feet leased. GSA has been requiring BIM for all major construction and modernization projects receiving design funding to be sufficient to support spatial program validation. GSA is developing guidelines for additional BIM capabilities in future projects.

One of the challenges related to BIM is pulling all the existing information together for the specific building being developed or used. In order to address this interoperability challenge, there have been several standardization efforts, such as the National CAD Standard guidelines, for uniformly organizing and presenting facility drawing information that streamlines the exchange of building design and construction data in drawings (buildingSmart Alliance 2008).

There are also efforts being developed for product modeling exchange, such as the Industry Foundations Classes (IFC), developed by the International Alliance for

Interoperability (IAI 2009), now known as buildingSMART International. IFC provide a formalized representation of typical building components (i.e., wall, door), attributes (i.e., type, function, geometric description), relationships (i.e., physical relationships, such as supported-by, connected-to), and more abstract concepts, such as schedules, activities, spaces, and construction costs, in the form of entities. IFC are the most notable and widely accepted data models for buildings and they aim at enabling information exchange in the AEC/FM industries. IFC specifications contain a digital information structure of the objects making up a building, capturing the form, behavior, and relation of the parts and assemblies within the building (IAI 2007). In contrast to exchanging plans via drawing files, such as DXF or DWG, IFC exchange is strictly model based. A wall is not a set of lines, but an object with specified attributes and relations (Clemen and Grundig 2006). Each entity is represented as a class; thus each can have a number of properties, such as name, geometry and materials, relationships, and constraints on the relationships. Such a standard enables the use of semantics and parametric modeling, and supports information exchange in the form of models, as well as the use of these models to support more complex tasks in computer-aided design and construction.

In summary, computer-aided design has been around since the 1960s with efforts varying from computer-aided drafting, mimicking a drafter's manual work, to parametric and object-oriented CAD, which added intelligence and automation to design tasks. For more than a decade, the notion of semantics has also been added on top of objects in Building Information Models, providing even more power to the term computer-aided design. It helps to perform complex tasks, such as building energy performance simulations, schedule analysis through 4D simulations, and design coordination. Interoperability between different software systems is still an issue, but, as is discussed in this chapter, there have been many efforts that are leading toward having a more integrated modeling environment, enabling more effective collaboration and sharing of information between different stakeholders throughout facilities' lifecycle. The next section provides an overview of a wide variety of data standards developed for enabling exchange of information in early CAD applications to exchange of semantically rich building information.

4.3 Overview of Data Standards and Specifications Utilized within the AEC/FM Domain

Many data standards and specifications have been developed for seamless data exchange between multiple applications within the AEC/FM industry. Efforts to bring standardization into data exchange and transfer between parties can be grouped based on their coverage in representing and exchanging AEC/FM related information as (1) early efforts targeting exchange of geometry and topology data only, (2) aspect model exchange standards and specifications, (3) building information model exchange standards and specifications, and (4) other standards or specifications that can assist in exchanging information needed in the AEC/FM industry.

The data standards and specifications introduced in this section are used to exchange information on products and processes within the AEC/FM domain at different stages of a project. Table 4.1 provides an overview of existing standards in terms of the agency leading the development effort, the year in which the development was initiated, the project phases which the specific data standard target to improve interoperability, semantics represented, the areas of use, typical file format used for data exchange, and the possibility/mechanisms to extend the existing versions of the data standards.

4.3.1 Early Standards and Specifications Targeting Geometry and Topology Information/Data Only

Initial efforts for maintaining interoperability within the AEC/FM industry were focused around developing neutral file formats, within which geometric and some topologic information are depicted. Utilization of neutral data formats required translators from specific software applications to neutral file formats. Drawing eXchange Format (DXF), Initial Graphics Exchange Specification (IGES), and STandard for the Exchange of Product model data (STEP) are three data exchange efforts that were developed in the 1980s and have been predominantly used within the AEC/FM industry.

4.3.1.1 Drawing eXchange Format (DXF)

DXF was initially developed by Autodesk for enabling interoperability between CAD applications (Eastman 1999). It was launched as part of AutoCAD 1.0 in 1982. It supports ASCII and binary formats, and is used especially to exchange 2D geometrical data (points, lines, arcs, polygons, text) of entities represented in various CAD packages developed by Autodesk. A ".dxf" file includes information about a drawing in various sections (Autodesk 2008): (1) *header section*, which provides information about variables associated with a drawing; (2) *classes section,* which provides information about application specific classes, whose instances appear in the other sections of the file; (3) *tables section*, which contains a set of tables, each of which provides definitions of used terms; (4) *blocks section*, which contains information about entities that are used to define each block; (5) *entities section*, which contains information about graphical objects; and (6) *objects section*, which contains information about entities that have no graphical or geometrical meaning. Advantages of DXF include small file sizes and efficient exchange of 2D graphical data.

Though DXF enables sharing of 2D geometric information, there are limitations of DXF in terms of its capability in supporting semantically rich data exchange.

Table 4.1   A Summary of Data Standards Utilized within the AEC/FM Industry

Data Spec/ Standard	Development Group & Project Starting Year	Targeted Project Phases	Semantics	Usage	File Format	Extensibility
Early Standards Targeting Geometry and Topology Information/Data						
IGES	Boeing/ General Electric (1979)	Design	2D/3D geometries, topological relations, non-geometric data	Exchange 2D/3D CAD data among various CAD applications	IGES based on ASCII	By development team
DXF	Autodesk (1982)	Design	2D geometry	Exchange CAD data among various CAD applications	DXF based on ASCII	By development team
STEP	ISO technical committee 184/ SC4 (1984)	Design, Fabrication, Erection	2D/3D geometries, topological relations, non-geometric data	Exchange 2D/3D CAD product data throughout their lifecycle	STEP Part 21	By development team

Product Model Exchange Standards/Aspect Models

CIS/2	University of Leeds and AISC (1995) as STEP Application Protocol 230	Design, Analysis, Fabrication	Geometry, location, orientation, parts, assemblies, bolts, holes, welds, sequences, materials, surface treatment, connections, properties	Exchange of structural steel design, analysis, and fabrication information	STEP Part 21	Consensus among the software implementers and AISC
gbXML	Green Building Studio (2000)	Design	Geometry, spatial, geography (building coordinates)	Exchange building CAD model to building simulation models	XML	XML extension mechanisms
bcXML	eConstruct group (2000)	Design, Construction, Supply Chain, Operations and Maintenance	Products, properties, taxonomy of terms and language rules	Exchange of construction products, resources, work methods, regulations	XML	XML extension mechanisms

(continued)

Table 4.1 A Summary of Data Standards Utilized within the AEC/FM Industry (Continued)

Data Spec/ Standard	Development Group & Project Starting Year	Targeted Project Phases	Semantics	Usage	File Format	Extensibility
Product Model Exchange Standards/Aspect Models						
IFC-Bridge	IAI-France (2002) IAI-Bridge (2007)	Design, Construction	Bridge physical elements, geometry, parts, location, and property information	Exchange of bridge structure information	STEP Part 21, ifcXML	XML extension mechanisms
OBIX	OASIS (2003)	Facilities/OM	Sensor information (value, range, status) exchanged between building automation systems	Exchange of sensor information between building automation systems	XML and Web services	XML extension mechanisms
AEX	FIATECH (2004)	Design, Supply Chain, Facilities/OM	Product information of equipment, properties (e.g., material), document associated with equipment	Exchange engineered equipment information	XML	XML extension mechanisms

agcXML	NBIMS Committee (2006)	Design, Construction Supply Chain, Operations and Maintenance	Documents	Exchange of construction-related business-to-business documents	XML	XML extension mechanisms
BPC	FIATECH (2006)	Design, Manufacturing, Installation	Precast concrete members, parts, geometry, location and connection information	Exchange of information for design, manufacturing and installation of precast concrete members	-	-
Building Information Modeling Standards						
IFD	International Construction Information Society (1995)	Design, Construction, Supply chain, Facilities/OM	Products as concepts, properties, units and values, relationships between these concepts	Exchange of language and application independent description of construction products	XML	Through defining new concepts

(continued)

Table 4.1   A Summary of Data Standards Utilized within the AEC/FM Industry (Continued)

Data Spec/ Standard	Development Group & Project Starting Year	Targeted Project Phases	Semantics	Usage	File Format	Extensibility
Building Information Modeling Standards						
IFC	IAI, also known as BuildingSmart (1996)	Design, Construction, Supply Chain, Facilities/OM	Products and associated elements, geometry, properties, geography, topology, relationships, cost, schedule, people, organization, site, documents	Exchange project information (product, process, control)	STEP Part 21, XML	Formal extension mechanisms
NBIMS	NIBS (2005)	Design, Construction, Supply Chain, Facilities/OM	Information required about all aspects of a facility throughout its lifecycle	Exchange of information about facilities throughout their lifecycle	Refers to other standards	Through model view definition diagrams

Other Data Standards

Sensor ML	Open Geospatial Consortium and Sensor Web Enablement Working Group (1998)	Construction, Facilities/OM, Disaster Management	Metadata for identification, classification, description, constraints, history, capabilities, and accuracies of sensors, physical objects and non-physical objects associated with sensor systems, lineage of observations, interconnections between sensors, post-processed data	Exchange of in-situ or remote sensor identification, location and observation data of sensors	SML	Major elements are fixed, but can be extended when needed
landXML	Industry consortium initiated by Autodesk (2000)	Surveying, Urban Planning, Disaster Management	Elements (address point, boundaries, daily traffic volume, parcel), complex types (point type, raw observation type) and simple types (angle, area, slope, speed, zone surface type)	Exchange of data created during the land planning and land surveying processes	XML	XML extension mechanisms

(continued)

Table 4.1 A Summary of Data Standards Utilized within the AEC/FM Industry (Continued)

Data Spec/ Standard	Development Group & Project Starting Year	Targeted Project Phases	Semantics	Usage	File Format	Extensibility
Other Data Standards						
KML	Open Geospatial Consortium and Standards Working Group (2001)	Environmental, Urban Planning Disaster Management	Geometry elements derived from GML 2.1.2, including point, line string, linear ring, and polygon	Exchange of geographic visualization information including annotation of maps and images	XML	XML extension mechanisms
cityGML	CityGML1.0 Standards Working Group (2002)	Design, Construction, Facilities/OM, Urban Planning, Disaster Management	Geometrical, spatial, topological, properties	Exchange of mainly spatial information for urban and landscape planning, disaster management, homeland security	XML	XML extension mechanisms

IFG	Norwegian Planning Authority (2003)	Design, Surveying, Construction	Geometrical (maps, contours, coordinate systems), properties, spatial information (building story and individual space)	Exchange of mainly geographical information	XML STEP21	XML extension mechanisms

* project start year shows the year of initiation for developing the specifications.

DXF does not contain topology information, nor does it process all entity attributes. In addition, available DXF exchange processors are not generic in converting 2D entities (e.g., the same entities can be converted as polylines or line segments) and this results in trial-and-error of users to pick and use the translator that best serves their needs (Eastman 1999). These result in interoperability problems, especially in sharing 3D geometric information.

4.3.1.2 Initial Graphics Exchange Specification (IGES)

IGES is another application-independent neutral file format developed for the exchange of CAD data. It was initiated in 1979 by Boeing and General Electric and then accepted as a standard (ANSI Y14.26M) by ANSI in 1981. IGES is capable of exchanging: (1) 2D/3D geometries, such as curves and surfaces; (2) topological relations, such as connectivity between geometric entities; and (3) some non-geometric data, such as properties of entities, dimensions and drafting notations (US PRO 1996). IGES uses ASCII file format and is composed of various sections as a DXF file. It starts with an optional *flag section*, which defines whether the file is in binary or compressed format. A *start section* follows the flag section, and it provides a description of the contents of a file. Following that, a *global section* provides information needed for pre- and post-processors. The rest of an IGES file contains a *directory entry section*, which keeps an index of the file and attribute information for each entry; *a parameter data section*, which contains entity type numbers, pointers to entities, and pointers to attributes in tables; and finally a *terminate section* that shows the end of file (US PRO 1996).

Though IGES is an early vendor-neutral data exchange standard and is capable of exchanging geometric and topological information, it has certain limitations. IGES does not describe non-geometric information about a model. In addition, CAD systems require a translator to read the original file format. The utilization of translators might result in describing the same geometric entities in multiple ways (e.g., boundary representation vs. swept solid), and incorrect mapping of data exported from a CAD system to IGES representation, resulting in unrecognized entities in the postprocessors; hence, translators might reduce the quality of the model. Also, IGES can result in large file sizes and require long processing times (Slansky 2005).

4.3.1.3 STandard for the Exchange of Product model data (STEP)

STEP is a data standard being developed by ISO technical committee 184 subcommittee SC4 since 1984 as an international standard (ISO 10303) for exchanging 3D product data. STEP can help in storing product data archives throughout a product's lifecycle and exchange product data in a neutral format. In addition to 3D geometric representation of any type of product (e.g., a building, a steel structure), STEP supports exchanging topology (e.g., edge, vertex), tolerances, assemblies, configuration, and attribute (e.g., surface finishes, material properties) information. STEP also uses ASCII file format and the standard consists of several parts, including (Eastman 1999): (1) *description methods*, which provide information about which modeling language, such as EXPRESS, is being used to model information in integrated resources and application protocols; (2) *integrated resources*, which provide information to represent a common single definition of product information as applications change; (3) *application protocols*, which specify scope and requirements of a domain-specific application for the data model; (4) *implementation methods*, which define resources for STEP implementation, such as STEP physical file and data access interface; and (5) *conformance testing*, which assesses whether STEP languages and files, such as EXPRESS, are used and implemented properly. STEP has advantages over IGES and DXF, as it focuses on a product data model for the domain semantics, then specifies the data format. This data model includes data items related to topology, properties and assemblies, and targets incorporating data not only from the design phase, but also from later phases, such as operations and maintenance of a facility, depending on the application protocols.

Limitations of STEP exchange include large and complex documentation, time-consuming development of STEP translators, and large file sizes due to a large number of objects represented from whole product lifecycles (Ball et al. 2007, Slansky 2005). The limitations of early geometry and topology exchange standards led to continuous efforts for enabling interoperability within the AEC/FM industry. The next section provides an overview of the exchange standards developed for exchanging domain specific information within the AEC/FM industry.

4.3.2 Product Model Exchange Standards and Specifications/Aspect Models

Many data exchange specifications and standards/data models have been developed since the late 1990s for enabling the exchange of information items associated with a specific domain or phase of a project. These data standards, which were developed to address data exchange requirements of specific domains or operational level departments, are also referred to as aspect models (Eastman 1999). These standards and specifications include the following:

1. CIMsteel Integration Standards (CIS/2): This specification was developed for enabling exchange of structural steel design, analysis, and fabrication information (Crowley and Watson 2000).
2. Green Building eXtensible Markup schema (gbXML): This specification aims at enabling exchange of design, certification, operation, and maintenance information for resource efficient buildings (gbXML 2008).
3. Building Construction eXtensible Markup Language taxonomy (bcXML): This standard was developed for enabling exchange of construction terms, definitions, properties, units, names in different languages and alphabets (Rees et al. 2002).

4. Industry Foundation Classes for Bridges (IFC- Bridge): This standard was developed for enabling exchange of bridge engineering information as an extension to the existing IFC standard (Yabuki and Li 2006).

5. Open Building Information eXchange (OBIX): This standard is being developed for enabling exchange of information for having intelligent buildings with the target of having integration for technologies utilized for security, HVAC, and building automation (OBIX 2008).

6. Automated Equipment Information eXchange (AEX): This set of XML schemas was developed as a specification for enabling exchange of equipment design, procurement, delivery, operation, and maintenance information (FIATECH 2008).

7. Associated General Contractors of America XML (agcXML): This set of XML schemas was developed as a specification for enabling exchange of construction-related business-to-business data that are currently exchanged on paper documents (Tardiff 2007).

8. BIM for Precast Concrete (BPC): This specification was developed for enabling exchange of design information for precast concrete components (Eastman et al. 2008).

An overview for each data standard/specification is provided in the following subsections.

4.3.2.1 CIMsteel Integration Standards (CIS/2)

CIS/2 is the product data model for structural steel that facilitates the exchange of information between steel design, analysis, detailing, and fabrication software (Crowley and Watson 2000). It was developed as a research project at the University of Leeds, as part of the Pan-European Eureka CIMsteel project (CIS 1997), and was adopted by the American Institute of Steel Construction (AISC) in 1998 as their data exchange format for interoperability between steel-related software applications. CIS/2 has been widely implemented in many steel specific software packages and in some general-purpose BIM software.

CIS/2 uses some of the STEP resource models and supports three different views or models of structural steel: design, analysis, and manufacturing. The manufacturing model is also known as a physical, detailed, or fabrication model. There is a logical relationship between the three models. For example, a beam that is subdivided into several elements for analysis is logically only one beam in a design or manufacturing model. A connection in a design model that only indicates that two parts are connected to each other is logically, in a manufacturing model, a fully detailed connection with bolts, holes, welds, and gusset plates.

The CIS/2 analysis model represents steel structures as analysis nodes, elements, loads, reactions, and boundary conditions. Prismatic parts in the design and manufacturing model are defined by a cross section designator, length, position, and orientation. Curved parts, flat and bent plates, and corrugated decking can also be modeled, as can connection materials, such as bolts, holes, and welds. Parts can also be grouped into assemblies and sequences and assigned surface treatments, material grades, and functional characteristics.

A mapping has also been developed between CIS/2 and the Industry Foundation Classes (IFC) used in the general building industry for information exchange (Lipman 2009). In some cases, there is a direct one-to-one mapping between CIS/2 and IFC entities and concepts, while in other cases there is a one-to-many or one-to-none mapping. The mapping shows that while IFC can easily model, with multiple representations, the geometry of steel structures, some of the semantics in CIS/2 have no equivalent in IFC. For example, the geometry of bolts can be modeled, in IFC; however, there is no concept that the bolts are in a specified pattern as there is in CIS/2. The mapping has pointed out other deficiencies for modeling structural steel in IFC. The mapping is implemented as a translator from CIS/2 files to IFC files.

4.3.2.2 Green Building Extensible Markup Language (gbXML)

gbXML specification is a data model developed for exchanging files or messages associated with exporting CAD model information of a facility to design and energy consumption simulation tools. It is an effort led by Green Building Studio with support of the California Energy Commission's Public Interest Energy Research (PIER) Program, and the California Utilities Companies since 2000. It is based on extensible markup language (XML) to enable sharing data with other applications.

Information which can be represented with gbXML, includes building information for space, surfaces and zones; surface types; space area and air volumes; building type; building geographic coordinates; and information for light fixture elements. Table 4.1 provides an overview of this data standard.

gbXML was specifically designed for building energy simulation. In addition, information such as material U-values, space occupancy schedule, and global building coordinates generated from building simulation tools cannot be imported back to original applications with the added information. gbXML is currently utilized for solving re-entering or reformatting spatial and geometric data used by building energy simulation tools.

4.3.2.3 Building and Construction Extensible Markup Language (bcXML)

Building and Construction Extensible Markup Language (bcXML) is a taxonomy of terms and language rules developed for enabling exchange of construction product, resource, work method, and regulation information for the e-business communication process (Rees et al. 2002). It was developed within the eConstruct project in 2000. bcXML can represent names, definitions of objects (concepts) and relationships between them, properties, and measures of properties that are related to building construction projects.

Similar efforts have developed taxonomies for exchanging product information, such as LexiCON of the Netherlands, and Barbi of Norway (Lima et al. 2007). Mappings between these data models exist, such as translators between LexiCON and bcXML. These European-based efforts have resulted in the development of an international standard (i.e., IFD).

4.3.2.4 Industry Foundation Classes for Bridges (IFC-Bridge)

IFC-Bridge data standard is being developed for enabling the exchange of bridge engineering information as an extension to the IFC standard. The roots of this data model come from two separate research studies from France and Japan. In 2002, these two groups were joined, with the support of IAI, to develop the IFC-Bridge standard (Yabuki and Li 2006).

The IFC-Bridge data model includes information about the general structure of bridges; complete geometry information about bridge spatial elements, physical elements, and element parts; material properties, pre-stressing, and process control (Arthaud and Lebegue 2007). This data model has been developed as an extension to the IFC schema similar to BIM for precast concrete, in order to detail information exchange for bridge components. The only limitation of this data model is the same limitation that comes with any domain-specific data model. It can only enable exchanging bridge-specific information items and need to be used hand-in-hand with industry-wide data standards.

4.3.2.5 Open Building Information Exchange (OBIX)

OBIX data standard is being developed for enabling the exchange of information coming from embedded sensors that sense information for various tasks, such as security, utilities, access control, lighting, and HVAC. It has been under development by the Organization for the Advancement of Structured Information Standards (OASIS) since 2003. It is based on XML and Web-services. This data standard enables communication between mechanical and electrical building control systems and front-end applications (OBIX 2008).

OBIX is currently capable of representing information items, such as objects (e.g., switches, lights), as well as references to URIs used to identify objects, status, and values in its object model. The main advantage of using this data model will be having a standard for exchanging information shared between various building automation systems, which currently rely on binary protocols (e.g., BACnet, LonTalk), that may experience problems with routers or firewalls as they are used over TCP/IP networks (OBIX 2008).

4.3.2.6 Automating Equipment Information Exchange (AEX)

AEX is composed of a set of XML schemas developed for exchanging equipment information in design, procurement, delivery, operations, and maintenance phases of a facility (FIATECH 2008). It is an effort that has been led by FIATECH, equipment manufacturers, software suppliers, industry associations, and NIST since 2004. Semantics represented within AEX include equipment information found on various equipment lists and bill of materials documents, process materials, associated properties, calculation methods, and experimental property data.

This data specification targets streamlining the flow in the equipment supply chain by enabling information exchange from design to equipment delivery. AEX specification is continuously evolving and currently covers centrifugal pumps, centrifugal fans, centrifugal compressors, reciprocating compressors, electric motors, air-cooled heat exchangers, shell and tube heat exchangers, control valves, and numerous other types of valves (FIATECH 2008). Semantic mapping studies were conducted to map AEX information to and from IFCs and the American Society of Heating, Refrigerating and Air-conditioning Engineers' (ASHRAE) data models (Begley et al. 2005). The Hydraulic Institute (pump manufacturers and suppliers) adopted AEX as the basis for their data exchange standard HI 50.7 and advanced AEX as the recommended data exchange standard for ISO 13709.

4.3.2.7 The Associated General Contractors of America (AGC) Extensible Markup Language (agcXML)

agcXML was developed to enable the exchange of transactional information that parties, such as architects, engineers, and suppliers within the building construction domain, exchange (Tardif 2007). This is an effort led by the Associated General Contractors of America (AGC) and the National Institute of Building Sciences since 2006. It can be used to represent construction-related business-to-business data that is exchanged in documents, such as owner/prime contractor agreements, owner/construction manager agreements, contractor/subcontractor agreements, schedules of values, and requests for information.

4.3.2.8 BIM for Precast Concrete (BPC)

BPC data schema is a data model developed for enabling the exchange of design information for precast concrete components. It is an outcome of a project initiated in 2006 by the Fully Integrated and Automated Technology (FIATECH) consortium and led by a research team composed of an architecture firm, 3D precast companies, academicians, NIBS, and FIATECH.

Targeting interoperability between architects and precast contractors, BPC suggests extensions to IFC 2x3 by identifying information items that are exchanged specifically for precast elements. These information items include geometry of precast components, their details, and properties needed during design, fabrication, and erection (Eastman et al. 2008). This data model focuses on information exchange for the precast concrete domain and works hand-in-hand with IFCs.

4.3.3 Semantically-Rich Building Information Model Exchange Standards and Specifications

While many of the domain-specific aspect models discussed above are useful and successful within the specific disciplines that they are targeting, there are still cross-domain and cross-discipline data exchange needs and interoperability issues that need to be addressed. The specifications described in this section target these needs. Due to the fact that such specifications are more difficult to develop, these are a smaller number of standards of this type. Currently available standards include (1) Industry Foundation Classes (IFC), developed for enabling the exchange of facility-related information throughout its lifecycle; (2) International Framework for Dictionaries (IFD), developed for enabling the exchange of AEC/FM-related products' definitions, properties, units, values and relationships between products; and (3) the new effort to develop the National Building Information Modeling Standard (NBIMS), aimed at developing an integrated lifecycle information model, based on existing open standards.

4.3.3.1 International Framework for Dictionaries (IFD)

IFD is an ontology that is being developed to exchange construction product information in multiple languages (Bjorkhaug and Bell 2007). The International Construction Information Society and ISO TC 59/SC 13/WG 6 have been leading this effort since 1999. IFD can represent products as concepts, their properties, units and values, and relationships between these concepts. The IFD data model provides product specific information, such as what it is; what parts, properties, measures and values it has; and which will be required at different phases of a project. IFD provides this ability by defining a controlled vocabulary of names of objects. With this ability, IFD provides a bridge between building information models (e.g., IFC-based) and databases that contain product specific information (buildingSMART 2008). Industry foundation classes (IFC) can define to a level, the components, relations between components, and properties; whereas IFD can provide information about what each component is about, such as its global ID, measuring units, material definitions, name, and descriptions in a multiple languages.

4.3.3.2 Industry Foundation Classes (IFC)

IFC represent a specification for exchanging and sharing information throughout the life-cycle of facilities. This specification has been being developed since 1996 by buildingSMART International (formerly known as the International Alliance of Interoperability [IAI]). IFC-based information can be exchanged using an XML or STEP 21 file format. A STEP Part 21 file is an ASCII file and is composed of a header and data sections within which every entity is stated with a unique number. ifcXML uses XML to exchange information contained in IFC and involves a conversion of IFC schema from EXPRESS representation to one based on XML (Nisbet and Liebich, 2005). IFC began with IFC 1.0, and currently IFC 2×4 is under development.

IFC was developed to enable the exchange of information about all aspects of a facility for all phases of a project from design to operations. In order to claim IFC compliance, software vendors must undergo a certification process with buildingSMART's Implementer Support Group, who test and certify a vendor's IFC implementation. Currently, IFC compliant software cover a wide area of AEC/FM domains, such as: design, structural engineering, HVAC design, thermal analysis, code checking, quantity take-off, and cost estimation.

IFC can enable the exchange of product information, such as walls, and columns, with their geometric representations and properties. It also defines topology (element connectivity, schematic design), relations between component and spaces, and spatial structures. Moreover, IFC incorporates non-product information, such as costs, schedules, resources, and documents. Each entity is represented as a class, thus each can have a number of properties such as name, geometry, materials, and relationships. Its latest release, IFC 2X3, has a total of 653 entity definitions. The capability of extension is provided by the IFC Property Sets. Shared product information can be from nine different domains, which are: HVAC, building controls, electrical, plumbing and fire protection, architectural, structural elements, construction management, structural analysis, and facilities management.

The main architecture of the IFC model is divided into four layers: domain, interoperability, core, and resource. Each layer comprises diverse categories, and it is within each category or schema that the individual entities are defined.

Currently, IFC is the data model that has the widest scope for enabling interoperability within the AEC/FM industry. With its extensible representation, IFC is growing as more specific data exchange is needed for new design, construction, manufacturing, and operations tasks within the AEC/FM industry.

4.3.3.3 National Building Information Model Standard (NBIMS)

NBIMS is being developed to enable an integrated lifecycle information model, based on existing open standards. It has been a project of the NIBS since 2005. With the objective of having integrated lifecycle information, this effort targets developing standards of standards by defining the NBIMS requirement for interoperability. The main objective of NBIMS is to have standardized information about a facility by defining how facility information exchange should take place, what a building information model contains, and organizing facility lifecycle information (NBIMS 2007).

For this purpose, NBIMS considers a shared building information model at the center, and works to define requirements for a model to be considered a building information model. While doing so, NIBS works in close collaboration with various parties, such as designers, contactors, and software vendors to evaluate and extend existing industry-wide data standards, such as IFC and IFD.

4.3.4 Other Data Standards and Specifications

There are also data standards and specifications that facilitate the exchange of information and can be helpful within the AEC/FM industry, even though they were not originally developed for and within that industry. These data standards are related to exchanging information with GIS-based applications, and include (1) Sensor Model Language (SensorML) schema, which was developed for enabling the exchange of sensor-based information from different sensor applications; (2) Land eXtensible Markup Language (landXML), developed for enabling the exchange of data created during the land planning, civil engineering, and land survey process; (3) Keyhole Markup Language (KML) schema, which was developed for enabling the exchange of geographic visualization information including annotation of maps and images; (4) city Geographic Markup Language (cityGML), developed for enabling the exchange of geometrical, spatial, and topological data of water bodies, sites (currently building), transportation facilities, city furniture, generic city objects, and their properties; and (5) Industry Foundation Classes for GIS (IFG), developed for enabling the exchange of geographic information in GIS with the IFC schema. These data standards and specifications are examined in the following subsections.

4.3.4.1 Sensor Model Language (SensorML)

OpenGIS SensorML is a data model developed for providing a standard language to define sensor-systems and components that play a role in these systems associated with measurements, and post-processing of these measurements (OGC 2008). It was initiated by a group under a National Aeronautics and Space Administration (NASA) program in 1998, and has continued its development under the oversight of Open Geospatial Consortium (OGC), since 2000.

SensorML includes modeling sensors as processes that convert observable phenomena into observed values. It provides information for locating sensors, sensor observations, processing information from observations, and sensor properties (Botts 2002). Any discipline that needs sensor-based data/information can benefit from the utilization of the SensorML standard. Within the AEC/FM domain, SensorML can be used for exchanging information required during operations and maintenance (e.g., modeling different sensors for facility operations and management, navigation within facilities, security of facilities, maintaining occupancy comports); or during construction for progress monitoring. Hence, SensorML is helpful in the construction and post-construction phases of a project.

4.3.4.2 Land Extensible Markup Language (landXML)

landXML was developed to enable the exchange of data created during land planning, development, transportation, and land surveying processes (Cover 2004). It is a data schema under development since the beginning of 2000 by an industry consortium, initiated by Autodesk, and now comprised of 190 companies, government agencies, and universities. landXML can represent civil engineering and survey measurement data as elements (i.e., address point, boundaries, daily traffic volume, parcel), complex types (i.e., point type, raw observation type), and simple types (i.e., angle, area, slope, speed, zone surface type). It covers units, coordinate systems, design geometry data (including points, alignments, surfaces, lines, curves), roadways, pipe networks, plan features (e.g., fence lines, curbs), and survey observations.

Any discipline that needs exchanging of geospatial land information can use landXML. Within the AEC/FM domain, landXML can be used for exchanging information between civil/surveying CAD and geospatial software applications required during various tasks such as, site surveying, visualization during roadway design, road model generation, automated construction machine controlling, and infrastructure modeling (Crews 2006).

4.3.4.3 Keyhole Markup Language (KML)

KML is a data model developed for enabling exchange of geographic visualization information, including annotation of maps and images (OGC 2008). It is an effort initiated by Google and continued by Open Geospatial Consortium–Standards Working Group in 2001. It can be used to model and display geometric features (including points, line strings, linear rings, polygons and regions), models, images, and additional geospatial data such as coordinate systems, placemarks, and time stamps on 2D or 3D earth browsers, GIS applications, or mobile applications (OGC 2008).

Any discipline that needs exchanging and displaying and visualization of geographic information can use KML. Within the AEC/FM domain, KML can be used to help facilitate information exchange and visualization in various applications, such as CAD and GIS applications, and overlay information exchanged between these applications on earth browsers. For example, KML can be used to locate and visualize groundwater levels, existing utilities, or project sites.

4.3.4.4 City Geography Markup Language (cityGML)

cityGML is a data model developed for enabling the exchange of virtual 3D urban objects, such as buildings, bridges, water bodies, and construct city models (Groger et al. 2007). A group called "Special Interest Group 3D" in Germany and cityGML 1.0 Standards Working Group have been working on the development of this data model since 2002. The cityGML data model can represent geometrical, spatial, topological, and appearance (surface characteristics, texture, material) properties for buildings, vegetation, water bodies, sites (currently only building), transportation facilities, city furniture, and generic city objects.

cityGML can be used in many application areas, such as urban and landscape planning, architectural design, environmental simulations, and disaster management. This data standard can also be used to exchange data for applications within the AEC/FM industry. Example application areas are disaster simulation and mitigation, site surveying, land development, and planning.

4.3.4.5 Industry Foundation Classes for GIS (IFG)

Industry Foundation Classes for Geographic Information Systems (IFG) is a data model developed for enabling the exchange of geographic information in geographic information systems (GIS) with the IFC schema (IFG 2008). Since 2003, the Norwegian Planning Authority has beeen working on developing the IFG schema. It provides a bridge between IFC and standard geographic information exchange standards, such as geographic markup language (GML) (AEC3 2008). IFG can represent areas (land parcels), geometric representation of building elements, maps, contours, coordinate systems, networks, distribution systems (water, sewer, power), proximity, survey data, terrain, semantic identification of a building, and building elements (building, wall, window, door, opening).

The aim of IFG schema was to use existing capabilities of IFC in representing data items that are related to GIS applications and extend it as needed. So, the basic idea was to create an overlap between the data models used within the AEC/FM and GIS domains. For developing the IFG schema, developers explored the capabilities of IFC in representing (1) positioning of objects in coordinate systems (which IFC represents with the IfcCartesianPoint entity); (2) building services, such as pipes and cables and their identification (which IFC enables with the IfcSystem entity); (3) geographic features (where IFC was extended to have an IfcGeographicalElement as a subtype of the IfcProduct entity); (4) qualified geometry, where geometric information differentiation with unique identifiers is required in GIS applications (which IFC provides with an IfcAnnotation entity); (5) shape of terrains (which IFC represents using the IfcSite entity as a grid or a triangulated irregular network (TIN); (6) proximity information (for which IFC was extended to include the proximity relationship); and (7) spatial structure arrangements (which IFC represents with IfcBuilding and IfcSpace entities) (IFG 2008).

Applications of IFG in the AEC/FM domain are various. While IFG is mainly used to exchange information between GIS and CAD applications, it can be used to store geographic and building information using a single data representation, as well as to facilitate zone and building plan submission processes by enabling sharing of location maps, utility services and zoning information. Other applications can be fire response management, disaster management, and integrating subsurface infrastructures with building information models.

4.4 An Evaluation and Comparison of Reviewed Data Standards and Specifications

This section provides a comparison of the data standards and specifications discussed in the previous sections in terms of their capability to represent and exchange information items shared within the AEC/FM domain. In order to perform this evaluation and comparison, a list of semantic information groups was created based on an exploration of workflows occurring between a set of disciplines within the AEC/FM industry. Based on this semantics list, the data standards and specifications were clustered by the disciplines that require identified semantic information groups to be exchanged among them. The disciplines whose workflows were examined for this study and the identified list of semantics are listed in Table 4.2.

The disciplines considered for the evaluation of standards and specifications were selected such that they are representative of all groups involved in a construction project from inception to the operation/maintenance phases. Examples of these disciplines are design groups, suppliers, and urban planners, as detailed in Table 4.2. The identified groups of information items include (1) products and associated elements, (2) geometry, (3) spatial information, (4) properties, (5) geography, (6) topology, (7) relationships, (8) cost, (9) schedule, (10) people, organization, and site, and (11) documents. These groups of information items were identified based on the explorations of the available data standards and specifications in terms of what they could represent, and clustering the outcomes.

A comparison of different rows in Table 4.2 shows that most of the existing standards and specifications include information items related to "products and associated elements" and "properties" of these elements. This is not surprising, since in order to exchange information about specific products (e.g., walls, columns, bridge elements), they first need to be defined and represented with their associated properties. "The geometry" information group follows these two semantics groups, as being the most highly represented information group. In terms of the least represented groups of information items, even though a large number of disciplines needed to exchange information related the to "people, organization, and site," "cost," and "schedule" groups, only industry wide data standards, such as IFC, have incorporated those items in their specifications.

A comparison of different columns in Table 4.2 shows that the majority of the information items needed for designers, construction groups, and facility management groups was represented within a large number of data exchange standards. This is mainly due to the fact that these groups have been targeted early in the standardization efforts. As new aspect models are developed specifically targeting the information exchange needs of disciplines that are not widely represented, gaps will be filled.

A general observation about existing data exchange standards and specifications is that these deal with semantic information groups at different scales,

Table 4.2 Data Standards and Specifications That Can Represent Semantics, by Semantic Information Groups and AEC/FM Disciplines

Semantics/ Groups	Designer Groups	Surveyor Groups	Contractor Groups	Supplier Groups	Subcontractor Groups	Facilities/ OM/Owner Groups	Environ- mental Groups	Urban/ Disaster Mgmt. Groups
Documents	AEX, AGCXML, IFC		AGCXML, IFC	AEX, AGCXML	AGCXML, IFC	AEX, AGCXML, IFC		
People/ Org./Site	IFC, IFG	IFG	IFC, IFG		IFC	IFC		
Schedule	IFC		IFC	IFC	IFC	IFC		
Cost	IFC		IFC		IFC	IFC		
Relation- ships	BPC, CIS/2, IFC, IFD		BPC, CIS/2, IFC IFD, SENSORML,	BPC	BPC, CIS/2, IFC, IFD	BPC, IFC, IFD, SENSORML		
Topology	BPC, CIS/2, CITYGML, IFC, IFG	IFG	BPC, IFC, IFG	BPC, IFC, IFG	BPC, IFC, IFG	CITYGML, IFC	CITYGML	CITYGML
Geography	CITYGML, GBXML, IFC, IFG, LANDXML	CITYGML, IFG, LANDXML	CITYGML, IFG, IFG, LANDXML		CITYGML, IFC, IFG, LANDXML	CITYGML, IFC, IFG, LANDXML	CITYGML IFG, KML, LANDXML	IFG, CITYGML, KML, LANDXML

Properties	AEX, BCXML, BPC, CIS/2, CITYGML, IFC, IFD, IFG	IFG, LANDXML	BCXML, CITYGML. IFC, IFD, IFG, SENSORML	AEX, BCXML, BPC, IFD	BCXML, IFC, IFD	AEX, BCXML, BPC, CITYGML, IFC, IFD, OBIX, SENSORML	CITYGML	CITYGML, KML, LANDXML
Spatial	CITYGML, GBXML, IFC, IFG, KML, LANDXML		CITYGML GBXML, IFC, IFG, KML, LANDXML		GBXML, IFC, IFG, KML, LANDXML	GBXML, CITYGML, IFC, KML, LANDXML	CITYGML, IFG, KML, LANDXML	CITYGML, IFG, KML, LANDXML
Geometry	BPC, CIS/2, GBXML, IFC, IFG, IFC-BRIDGE	CITYGML, IFG, KML, LANDXML	BPC, CIS/2, IFC, IFC-BRIDGE, IFG	CIS/2, CITYGML, IFC, IFG, KML, LANDXML	BPC, CIS/2, IFC, IFC-BRIDGE	BPC, CITYGML, IFC, GBXML, KML, LANDXML	CITYGML, KML, LANDXML	CITYGML, KML, LANDXML
Products, associated parts, connections	AEX, BPC, CIS/2, IFC, IFD, IFC-BRIDGE, IFG,	IFG, LANDXML	BCXML, BPC, CIS/2, CITYGML, IFC, IFC-BRIDGE, IFD, IFG, SENSORML	AEX, BCXML. CIS/2, IFD	BCXML, BPC, CIS/2, IFC, IFD, IFC-BRIDGE	AEX, BCXML, BPC, CITYGML, IFC, IFD, OBIX, SENSORML	CITYGML, KML	CITYGML, KML, LANDXML

for example, how a building site represented in IFC is different from how it is represented in IFG. Similarly, how a building is represented in IFC is different than how it is represented in cityGML. A question arises as to how these different scales can be integrated in standard representations, or how interoperability can be maintained between applications utilizing these different scaled representations. This issue will be more thoroughly discussed in other chapters in this book, where data standards integration issues, such as CAD/GIS integration issues, are discussed in detail.

4.5 A Discussion on the Current State of Interoperability and a Projection toward the Future

In an ideal world, exchanging information specified by a data standard between CAD applications would result in 100% of the expected information to be exchanged and accessible in the receiving CAD application 100% of the time. For some of the data exchange standards, it is known that this is not true for a variety of reasons. Some possible sources of the problems are (1) issues with mapping to and from internal software representations of information to the data exchange standards; (2) incomplete or incorrect implementations of the data exchange standard; (3) the data standard does not meet all the requirements of the information needed to be exchanged for a particular domain; and (4) inconsistent utilization of a software system, which results in information being mapped to the wrong data element in the exchange standard.

All data exchange standards should have some form of validation, conformance, and interoperability testing to be successful. Validation testing is the process to evaluate a standard to determine whether it satisfies the information exchange requirements of a particular domain. Conformance testing is the assessment of a software implementation in terms of whether it meets the requirements of a standard, i.e., is the software generating the correct information in the data exchange files (Kindrick 1996). Interoperability testing is the assessment of the end-to-end functionality between two software implementations.

Conformance testing involves developing specifications of what information is to be modeled or exchanged, creating the model or information in the software application that is being tested, generating the data exchange file, analyzing the file for correctness, and reading the data exchange file. The analysis of the data exchange file can usually be done with various software tools. Interoperability testing extends the process by importing the data exchange file into another software application and evaluating the resulting model or information in the receiving application. The evaluation compares the original representation in the first application to the resulting representation in the second application. Successful conformance testing leads to better assurance that interoperability testing will also be successful. Interoperability testing without conformance testing can lead to software modifications compensating for non-conformance to the data exchange standard. This leads to implementations that do not universally interoperate with other similar applications.

Some data exchange standards have rigorous definitions of testing. For example, the STEP data exchange standard defines an entire framework for methodology and a framework for doing conformance testing (ISO 10303 Parts 31–35). For the STEP application protocol AP 227 for exchange of spatial configuration information of process plants, a validation report (Kline 1996) summarizes the validation, conformance, and interoperability testing program. The report was created in conjunction with the development of the exchange standard so that a methodology was in place with a test suite and implementation guidance is provided to test software implementations while they are being developed. For the OpenGIS specification, software applications can validate their products through the Conformance and Interoperability Test and Evaluation Initiative (OGC 2002).

Leaders in the building industry are striving to use IFCs as a data exchange standard, which is essential for the successful implementation of BIM. While some BIM projects can take place all within one suite of software products that do not require the exchange of information with other software applications, many projects need to exchange information between different CAD and BIM software, and with applications such as energy analysis, quantity takeoff, and facility management programs. However, there have been several studies that point out various problems with exchanging information with IFC.

An interoperability test was carried out by the Danish chapter of the IAI (IAI Denmark 2006) that modeled a simple structure in five CAD applications, exported the model as an IFC file, and imported the file to the other four applications. A set of evaluation criteria was applied to the exported IFC files and to the resulting model in the CAD application. The testing and evaluation included aspects of conformance and interoperability testing. However, it was done on an ad hoc basis without reference to any established testing procedure. The test showed some elements missing in the resulting IFC files and CAD application models.

A benchmark test for interoperability for precast concrete data was part of a project related to BIM for Precast Concrete (Kaner et al. 2008, Eastman et al. 2008). The test specified a structure with representative precast structural elements that was modeled in several CAD applications, exported to IFC files, which were then imported to a different CAD application. The exported IFC files and resulting CAD models in the receiving CAD application were evaluated. The IFC files generated by different CAD systems varied greatly. These variations were caused by how the precast concrete elements were modeled in the CAD systems and how those elements were mapped to the IFC file. There were also some significant differences between the original and resulting CAD models such as objects with the wrong placement, missing elements, and geometry errors.

The ATC-75 project (ATC), which is developing IFC for structural components, also performed some ad hoc interoperability testing. The use case for the information exchange involved exchanging data from an architectural to a structural engineering model to do more detailed design. A benchmark test model of a section of a sports stadium was modeled in three CAD applications, exported to IFC, and imported to the other two CAD applications. The IFC files were evaluated by checking the file syntax and conformance to the IFC specification, and were visually inspected with IFC viewers. Discrepancies in the resulting CAD models were documented.

Several other projects have carried out testing through a comparison of IFC files (Palzar 2008, Ma 2006). The IFC files in those projects were generated in two ways. In the first scenario, given a representative model in a CAD system, an IFC file was exported and imported into a second CAD system. The second system then exported another IFC file which was compared to the original IFC. In the second scenario, the original IFC file was imported back into the original CAD application and a second IFC file was exported. This is commonly referred to as round-tripping. The original and the second IFC files are then compared. Each of the comparisons used different evaluation criteria. Comparing the IFC files to each other does give some measure of conformance and interoperability, but it does not take into account how the information might be modified when mapped multiple times to and from the CAD systems and IFC files. Comparing IFC files from different CAD systems is also not a representative workflow, particularly for round-tripping. The comparison of the CAD models is a more representative workflow.

All of the testing research projects described above have some aspects of conformance and interoperability testing. However, none of the testing was performed based on a rigorous methodology that: (1) defines how test models are specified to ensure coverage of the domain; (2) specifies how they are modeled in CAD applications; (3) ensures that a set of test models provides sufficient coverage for all data elements that need to be tested; (4) defines the verdict criteria that should be used to evaluate the resulting IFC file; and (5) specifies how the verdict criteria and testing process are used to evaluate the resulting CAD model and compare it to the original CAD model. The results of the tests are also only a snapshot in time of the state of interoperability. The tests were performed with specific versions of CAD software and IFC interfaces that most likely have been modified and upgraded since those tests took place. Results of the tests cannot necessarily be extrapolated to CAD software and IFC interfaces that are currently available.

The observations about the testing projects indicate the need for more well-defined, reliable, and repeatable testing methods for data exchange standards such as IFC. Such methods would be of great benefit to software developers for developing more dependable implementations of IFC information exchange, and for end-users to perform their own testing projects without having to reinvent the wheel and do it on an ad hoc basis.

4.6 Conclusions and Future Directions

Various data standards and specifications are being developed continuously for enabling interoperability within the AEC/FM industry. These range from early efforts (e.g., DXF, IGES) developed to exchange geometry and topology information, to product model data exchange (e.g., CIS/2, gbXML, STEP) and semantically-rich building information modeling exchange standards and specifications (e.g., IFC, IFD). Data standards and specifications were compared in terms of their ability to represent a set of information groups, such as products, properties, and geography. This groups were identified by combining and clustering the information items that could be represented by all standards. This comparison shows that the majority of the information items needed for various AEC/FM groups, such as designers, construction groups, and facility management groups are represented within a large number of data exchange standards. This is mainly due to the fact that these groups have been identified and hence targeted early in standardization efforts. In addition, it was observed that most of the existing standards include information items related to "products and associated elements" and "properties" of these elements.

Though the capabilities of these data standards in representing required information items by different disciplines are satisfying, there are still issues that hinder interoperability between applications. These issues were identified as (1) issues with mapping to and from internal software representations of information to the data exchange standards; (2) incomplete or incorrect implementations of the data exchange standard; (3) the data standards not meeting all the requirements of the information needed to be exchanged for a particular domain; and (4) inconsistent utilization of a software system that results in information being mapped to the wrong data element in the exchange standard. The main reason for the existence of such issues is due to the lack of well-defined, reliable, and repeatable testing methods to test the conformance, interoperability, and validation of the data standards and applications using these data standards.

As a future direction within the AEC/FM industry for solving the interoperability problems, there should be efforts to develop formalized, well-defined, reliable, and repeatable testing methods for deploying the developed data standards and specifications. In addition, multiple fragmented efforts need to be integrated so as to bring a true interoperable environment for the AEC/FM domain. Initial efforts for such large-scale integration of data standards are currently being performed. The National Building Information Modeling Standard (NBIMS), which is an ongoing effort led by the National Institute of Building Sciences, is one of these efforts.

Integration of data standards for enabling CAD and GIS integration is another example. Such efforts should incorporate a standard methodology for having conformance, interoperability, and validation tests for data standards. With that, true interoperable environments will be achieved within the AEC/FM domain without losing semantic integrity of information shared among applications and disciplines.

References

AEC3 (2008). "IFC for GIS-Norway." Available at: http://www.aec3.com /5/5_008_IFG. htm. Last accessed on January 12, 2009.

Arthaud, G., and Lebegue, E. (2007). "IFC-Bridge v2 data model: Edition R7." IAI French Chapter Report, February 2007. Available at: http://www.iai-tech.org/groups/ msg-members/projects/IFC-BRIDGE-V2-Data-Model-R7-draft.pdf/view. Last accessed on December 15, 2008.

ATC-75 Project: Development of Industry Foundation Classes (IFCs) for Structural Components. Available at: http://www.atcouncil.org/atc75.shtml.

Autodesk (2008). "DXF reference specifications." Autodesk 2009. Available at: http://usa. autodesk.com/adsk/servlet/item?siteID=123112&id=8446698. Last accessed on April 15, 2008.

Ball, A., Ding, L., and Patel, M. (2008). "An approach to accessing product data across system and software revisions." *Advanced Engineering Informatics* 22(2), 222–235.

Begley, E. F., Palmer, M. E., and Reed, K. A. (2005). "Semantic Mapping Between IAI ifcXML and FIATECH AEX Models for Centrifugal Pumps." NISTIR 7223.

Bjorkhaug, L., and Bell, H (2007). "IFD in a nutshell" SINTEF Building and Infrastructure. Available at: http://dev.ifd-library.org/index.php/Ifd:IFD_in_a_ Nutshell. Last accessed on April 14, 2008.

Bloor, M.S. and Owen, J (1995). *Product Data Exchange.* UCL Press: London.

Botts, M. (2002). "Sensor model language: SensorML for in-situ and remote sensors" OpenGIS Consortium Inc., Report #: OGC-02-26r1.

Bozdoc, M. (2004). "The history of CAD." Available at: http://mbinfo.mbdesign.net/ CAD-History.htm. Last accessed on December 26, 2008.

BuildingSmart (2006). "IFC for GIS

BuildingSmart (2008). "IFD library for buildingSmart." Available at: http://www.ifd-library. org/index.php/Main_Page. Last accessed on December 15, 2008.

BuildingSmart Alliance (2008) "United States national CAD standard." Available at: http:// www.buildingsmartalliance.org/ncs/. Last accessed on December 26, 2008.

CIS (1997). "CIS/1: CIMSteel integration standards." Available at: http://www.engineering.leeds. ac.uk/civil/research/cae/past/cis/cis.htm. Last accessed on February 12, 2009.

Clemen, C., and Grundig, L. (2006). "The industry foundation classes (IFC)—Ready for indoor cadastre?" XXIII FIG Congress. 1–9.

Cover, R. (2004). "LandXML." Available at: http://xml.coverpages.org/landXML.html. Last accessed on May 07, 2008.

Crews, N. (2006). "LandXML 1.1. general workshop: LandXML.org 2006." Available at: http://www.landxml.org/Workshops/Workshops/LandXML-1.1_2006/LandXML. org_2006.ppt. Last accessed on January 12, 2009.

Crowley, A., and Watson, A. (2000). "CIMsteel integration standards release 2." SCI-P-268, The Steel Construction Institute, Berkshire, England. Available at: http://www.steelbiz.org/Discovery/AllResults.aspx?q=%22P268%22. Last accessed on January 12, 2009.

Eastman, C. (1975). "The use of computers instead of drawings in building design." *Journal of the American Institute of Architects*. March: 46–50.

Eastman, C. (1999). *Building Product Models: Computer Environments Supporting Design and Construction*, Boca Raton, FL: CRC Press.

Eastman, C., Sacks, R., Jeung, Y., and Kaner, I. (2008). "BIM for architectural precast concrete." Final Report for the Charles Pankow Foundation, March 26, 2008.

FIATECH (2008) "Automating equipment information exchange (AEX)." Available at: http://www.fiatech.org/projects/idim/aex.htm. Last accessed on May 07, 2008.

Gallaher, M. P., O'Connor, A. C., Dettbarn, J. L., Jr., and Gilday, L. T. (2004) "Cost analysis of inadequate interoperability in the U.S. capital facilities industry." NIST GCR 04-867. 194.

GBS (2008). "Green building studio version 3.0 of GBS Web service." Available at: http://www.greenbuildingstudio.com/gbsinc/index.aspx. Last accessed on April 08, 2008.

GBXML (2008). "Green building XML schema: A building information modeling solution for our green world." Available at: http://www.gbxml.org. Last accessed on April 08, 2008.

General Services Administration (GSA) (2007). "3D-4D building information modeling." Available at: http://www.gsa.gov/bim. Last accessed on November 27, 2007.

GSA Office of Government wide Policy (2006). "Executive summary report: An overview of the U.S. Federal Government's real property assets." Federal Real Property Report, Fiscal Year 2005. 14p.

IAI: International Alliance for Interoperability (2009). Available at: http://www.iai-tech.org. Last accessed on February 18, 2008.

IFG (2008). "IFC for GIS." Available at: http://www.iai.no/ifg. Last accessed on April 15, 2008.

International Alliance for Interoperability (2009). "Industry Foundation Classes, Version 2X3." Available at: http://www.iai-tech.org/products/ifc_specification/ifc-releases/summary. Last accessed on February 18, 2009.

ISO 10303 Parts 31–35 Conformance Testing Methodology and Framework.

Kaner, I., Sacks, R., Kassian, W., and Quitt, T. (2008). "Case studies of BIM adoption for precast concrete design by mid-sized structural engineering firms." Information Technology in Construction, 13. Available at: http://www.itcon.org/cgi-bin/works/Show?2008_21. Last accessed on January 10, 2009.

Kindrick, J., Sauter, J., and Matthews, R. (1996). "Improving conformance and interoperability testing." StandardView 4(1). Available at: http://portal.acm.org/citation.cfm?id=230871.230883. Last accessed on January 10, 2009.

Kline, S., and Palmer, M. (1996). "Application Protocol 227 Validation Report Version 1.0." NIST Internal Report 5875, 1996. Available at : http://fire.nist.gov/bfrlpubs/build96/art130.html. Last accessed on January 10, 2009.

Lima, C., Zarli, A., and Storer, G. (2007). "Controlled vocabularies in the European construction sector: Evolution, current developments and future trends." *Complex Systems Concurrent Engineering*, Loureiro, G., and Curran, R., Eds. pp. 565–574, London: Springer.

Lipman, R. (2009). "Details of the mapping between the CIS/2 and IFC product data models for structural steel." *ITCON*, Vol.14, 1–13. Available at: http://www.itcon.org/cgi-bin/works/Show?2009_01.

Ma, H., Ha, K.M., Chung, C.K.J., and Amor, R. (2006). "Testing semantic interoperability." *Proceedings of the International Conference on Computing in Civil and Building Engineering*, Montreal, Canada, June 2006.

NBIMS (2007). "National building information modeling standard: Overview, principles and methodology." Version 1, Part 1. Available at: http://www.facilityinformation-council.org/bim/publications.php. Last accessed on March 12, 2008.

NIBS (2007). National Institute of Building Sciences. http://www.facilityinformationcoun-cil.org/bim/index.php. Last accessed Nov 2007.

Nisbet, N., and Liebich, T. (2005). "ifcXML implementation guide: version 1." IAI report, Available at: http://www.iai-tech.org/downloads/accompanying-documents/guide-lines/ifcXML%20Implementation%20Guide%20v2-0.pdf %20Implementation %20 Guide%20v1-0.pdf. Last accessed on December 15, 2008.

NIST: National Institute of Standards and Technology (2004). "Cost analysis of inadequate interoperability in the U.S. Capital Facilities Industry." NIST GCS 04-867. Available at: http://fire.nist.gov/bfrlpubs/build04/PD./b04022.pdf. Last accessed on April 08, 2008.

OBIX (2008). "Open building information eXchange: What is OBIX?" Available at: http://www.obix.org/default.htm. Last accessed on December 12, 2008.

OGC (2002). "Open Geospatial Consortium: Conformance and interoperability testing and evaluation, phase 1." Available at: http://www.opengeospatial.org /projects/initiatives/cite1. Last accessed on February 12, 2009.

OGC (2008) "Open geospatial consortium keyhole markup language." Available at: http://www.opengeospatial.org/standards/kml. Last accessed on May 07, 2008.

Pazlar, T., and Turk, Z. (2008). "Interoperability in practice: Geometric data exchange using the IFC standard." *Information Technology in Construction*, 13. Available at: http://www.itcon.org/cgi-bin/works/Show?2008_24. Last accessed on January 10, 2008.

Rees, R., Tolman, F., and Beheshti, R. (2002). "How bcxml handles construction semantics?" *CIB w78*, June 12–14, 2002, Aarhus School of Architecture, Aarhus, Denmark.

Slansky, D. (2005). "Interoperability and openness across PLM: Have we finally arrived?" ARC Advisory Group White Paper, Available at: http://www.arcweb.com. Last accessed on April 23, 2008.

Smith, D. (2007) "An introduction to BIM." *Journal of Building Information Modeling*, Fall 2007.

Sutherland, I. (2003). "Sketchpad: A man-machine graphical communication system" Technical Report No. 574, University of Cambridge, UCAM-CL-TR-574. Available at: http://www.cl.cam.ac.uk/techreports/UCAM-CL-TR-574.pdf. Last accessed on December 26, 2008.

Tardif, M. (2007). "AGC launches AGCXML project to help streamline the design and construction process." AIA Edges Newsletter, Fall 2007. Available at: http://www.aia.org/nwsltr_tap.cfm?pagename=tap_a_200611_AGCxml. Last accessed on April 16, 2008.

USPRO: the US Product Data Association (1996). "Initial graphics exchange specification 5.3." Available at: http://www.uspro.org/documents/IGES5-3_forDownload.pdf. Last accessed on April 23, 2008.

Yabuki, N., and Li, Z. (2006). "Development of new IFC-Bridge data model and a concrete bridge design system using multi-agents." Intelligent Data Engineering and Automated Learning, Lecture Notes in Computer Science. pp. 1259–1266.

Yezioro, A., Dong, B., and Leite, F. (2008). "An applied artificial intelligence approach towards assessing building performance simulation tools." *Energy and Buildings*, 40(4), 612–620.

Young, N.W., Jones, S.A., and Bernstein, H. M. (2007). "Interoperability in the construction industry: SmartMarket report." McGraw-Hill Construction, Available at: http://construction.ecnext.com/mcgraw_hill/includes/SMRI.pdf. Last accessed on April 14, 2008.

Chapter 5

Interoperable Methodologies in CAD-GIS Integration Standardization Efforts
The Open Geospatial Consortium Perspective

Carl Reed

Contents

5.1 Introduction and Background

In recent years, a great deal of technical innovation has been accomplished in the areas of computer-aided design (CAD), architecture engineering and construction (AEC), geospatial solutions, 3D visualization, and urban simulation. A variety of products, information, and services abound in each of these environments. A framework of data and service interoperability that can utilize these innovations and technologies should exist across the lifecycle of building and infrastructure investment: planning, design, construction, operation, and decommissioning. Such an integrated lifecycle information workflow is of interest to the geospatial community* because there is a growing need for technologies and information to effectively interoperate between these domains to support a range of vital services and decision support needs.

To put the issue in clear perspective, a 2004 report by the National Institute of Standards and Technology (NIST) titled "Cost Analysis of Inadequate Interoperability in the U.S. Capital Facilities Industry," [1] quantifies the annual cost of waste due to inadequate interoperability among computer-aided design, engineering, and software systems in the construction industry to be $15.8 billion. And this figure was just for the U.S. building industry.

Interoperability is identified as a critical aspect of CAD-GIS integration. For interoperability to work, many stakeholders have to agree on common semantics, interfaces, information models, schemas, and best practices that address specific

---

* The geospatial community encompasses the traditional GIS and Earth Observation technology domains as well as location services, location enabled sensor systems, and any other community that must enable location interoperability.

requirements. Such community agreements are typically documented as standards. A standard is typically a document, established by consensus and approved by a community of interest, that provides, for common and repeated use, rules, guidelines, or characteristics for activities or their results, aimed at the achievement of the optimum degree of order in a given context.

The Open Geospatial Consortium (OGC) is a standard development organization that collaborates with many other standards organizations. In the CAD-GIS context, the OGC collaborates with the National Institute of Building Sciences (NIBS) and the buildingSMART alliance, previously the International Alliance for Interoperability (IAI). These collaborations are described later in this chapter.

We begin with a short description of the OGC.

5.2 What Is the Open Geospatial Consortium?

The OGC® is organized as a tax-exempt "membership corporation," as defined in section 501(c)(6) of the U.S. tax code. The mission of the OGC is to promote the industry consensus development and use of advanced open systems standards and techniques in the area of geoprocessing and related information technologies. The ongoing standards work of the OGC is primarily supported by the volunteerism of our members. Membership and OGC interoperability test bed initiative management fees finance the Consortium activities, such as the communications and collaboration infrastructure, Web presence, and staff.

5.2.1 OGC and Standards Development

The primary product of the OGC is a "standard." An OGC standard is a document that details the engineering aspects (and rules) for implementing an interface or encoding that solves a specific geospatial interoperability problem. These documents are defined, discussed, tested, and approved by the members using a formal process. The reader should refer to http://www.opengeospatial.org/standards for information and downloads of the currently approved OpenGIS® standards.

At the same time, coordination and collaboration with other standards organizations and industry trade associations are critical to the success of the OGC standards work. Standards work, geospatial content, and service interoperability cannot stand in isolation. Therefore, the OGC maintains a number of close and very beneficial alliances with other standards organizations, including the International Standards Organization (ISO) and the Organization for the Advancement of Structured Information Standards (OASIS). Most importantly for CAD-GIS integration and the building industry, the OGC maintains a formal relationship with the International Alliance for Interoperability and the National Institute of Building Sciences (NIBS). The details of this relationship are provided in the Partnerships section of this chapter.

OGC standards can be found at the OGC Web site (http://www.opengeospa-tial.org). These standards are freely and publicly available, and can be downloaded from the Web site.

5.2.2 OGC Technical Committee—Where the Standards Work Is Done

The OGC Technical Committee (TC) uses a formal standard consensus discussion and approval process to define, approve, and maintain OGC standards. The Technical Committee is comprised of a number of working groups (WGs). These WGs provide a forum for discussion of key interoperability issue areas, discussion and review of standards, and presentations on key technology areas relevant to solving geospatial interoperability issues. The primary product of the TC is the processing and adoption of OGC standards (which are often drafted in OGC test beds). The TC is also responsible for the maintenance and revision of our adopted standards. The Technical Committee is organized to focus on both general and domain-specific standard development. The TC Policies and Procedures provide the rules and governance for the work of the TC.

Technical Committee Working Groups provide an open, collaborative forum for discussions, presentations, and recommendations on a variety of items and issues of interest to the members of that Working Group. Working Groups can work on domain specific interoperability issues, or on candidate standards or revisions to existing standards.

5.3 The OGC Member Response: The 3D Information Management (3DIM) Working Group

Using the framework described above, the OGC members determined that a special domain focus was required to address the issue of CAD-GIS integration interoperability. A focus group of members developed a charter. Upon approval by the OGC membership, the CAD-GIS Working Group was formed in 2005 to identify and act on opportunities to improve interoperability of geospatial data and services across the AEC, Building Construction, 3D, and GIS domains. In 2007, the working group changed its name to 3D Information Management (3DIM) to better reflect that the total scope of work is not just CAD-GIS, but the integration and use of geospatial content and services throughout the entire built lifecycle.

From the 3DIM charter:

> The 3D Information Management (3DIM) Working Group is facilitating the definition and consensus approval of new standards that enable infrastructure owners, builders, emergency responders, community planners, and the traveling public to better manage and share location based information about complex built environments. Effective sharing

and integration of geospatial data and services has eluded the geospatial and CAD industry for decades. Today, through the cooperation of diverse stakeholders, integrated infrastructure information systems will be achieved. OGC members and partners will work in an iterative development process to achieve incremental demonstrations of real solutions.

The 3DIM Working Group meets at face-to-face meetings four times a year. They also have a number of teleconferences between meetings to insure continuity of their activity.

5.4 Partnerships to Foster Collaboration in 3DIM Activities

The OGC, IAI, and NIBS have a formal Memorandum of Understanding* to cooperate in areas of interest for the 3DIM WG. OGC also has an additional MOU with IAI. The OGC and the IAI understand the critical benefits of promoting closer coordination and collaboration of our respective program activities to strengthen open standards-based interoperability between the building infrastructure, and the broader geospatial, Architecture Engineering Construction (AEC) and information technology (IT) communities.

The essential elements of these agreements are designed to foster stronger collaboration between the geospatial and building communities. The objectives as documented in the MOU, along with current status information, is as follows:

- Encourage joint member activities to advance work items of mutual interest. An initial focus was to assess member interest in advancing the creation of an ifcXML-IFC/GML model to improve the ability of IFC building models to be shared between the AEC and geospatial technologies and user communities. This is ongoing. Current successful activities are documented later in this chapter.
- Jointly investigate other potential areas of collaboration for standards development, standards harmonization, and outreach and education. Considerable progress has been made in this activity area. Please see a later section on OGC TestBed activities
- Identify, prepare, and disseminate outreach materials including journal articles, white papers, and related reference documentation, to improve understanding, support, and application of jointly addressed standards areas and projects. A number of outreach materials have been developed and deployed. More information on this joint activity is provided later in this chapter.

---

* A Memorandum of Understanding (MOU or MoU) is a document describing a bilateral or multilateral agreement between parties. It expresses a convergence of will between the parties, indicating an intended common line of action. (Wikipedia)

Of critical importance is that the MOU provides a mechanism for members of IAI to participate in the OGC 3DIM Working Group and test bed activities.

5.5 OGC Standards Activities Relevant to CAD-GIS-BIM

This section describes a number of existing OGC standards that have high utility in enhancing the ability to integrate traditional GIS content with engineering drawings, as-built drawings, surveys, sensors, and the many other forms of location referenced content required for use in the built environment value chain. These standards include the following:

- OGC and ISO Web Map Service (WMS) Interface Standard
- OGC Web Map Context Standard (WMC)
- OGC Web Feature Service (WFS) Interface Standard
- OGC and ISO Geography Markup Language (GML) Encoding Standard
- OGC CityGML Encoding Application Schema
- OGC KML 2.2 Encoding Standard

Each of these standards is now described in more detail

5.5.1 OpenGIS Web Map Service (WMS) Interface Standard

WMS was the very first OGC Web services standard. WMS was designed to solve a simple interoperability problem: providing a standard interface that allows any browser-based application to access and display geographic content from multiple distributed geospatial repositories where the content is typically stored in a variety of formats and structures. The first version of the WMS standard was approved in 1999. The most recent version [2] was approved as an OGC standard in 2006. WMS is also an ISO standard (ISO 19139:2006). There are currently thousands of WMS implementations providing access to over 500,000 map layers.

From the ISO document:

> A Web Map Service (WMS) produces maps of spatially referenced data dynamically from geographic information. This International Standard defines a "map" to be a portrayal of geographic information as a digital image file suitable for display on a computer screen. A map is not the data itself. WMS produced maps are generally rendered in a pictorial format such as PNG, GIF, or JPEG, or occasionally as vector-based graphical elements in Scalable Vector Graphics (SVG) or Web Computer Graphics Metafile (WebCGM) formats.

When two or more maps are produced with the same geographic parameters and output size, the results can be accurately overlaid to produce a composite map. The use of image formats that support transparent backgrounds (e.g., GIF or PNG) allows underlying maps to be visible. Furthermore, individual maps can be requested from different servers. The Web Map Service thus enables the creation of a network of distributed map servers from which clients can build customized maps.

5.5.2 OpenGIS Web Map Context (WMC) Version 1.1 Standard

The OpenGIS WMC Standard defines a standard way to create and use documents that unambiguously describe the state, or "Context," of a WMS Client application in a manner that is independent of a particular client and that might be utilized by different clients to recreate the application state. This specification defines the encoding for the Context using eXtensible Markup Language [XML 1.0]. The XML context document can then be saved into persistent storage and be recalled at a later date. The context information can then be used to recreate Web maps created using WMS in a previous session. The WMC documents can also be used to share presentations created using WMS.

The Context document provides both layer independent context information and layer specific information. General information elements are the bounding box in units of a particular Coordinate Reference System that represent the geographic extent of the map and a dimension as a pair of integers that represents the suggested pixel size of the map. This element also contains some metadata about this particular WMS Context such as a title, abstract, keywords, and other information.

The layer specific elements encapsulate all the layers in the current context. These are essentially the WMS http requests. One or more of those layers may be retained in the context but can be hidden from the display.

5.5.3 Open GIS Web Feature Service (WFS) Interface Standard

The Web Feature Service (WFS) Implementation Standard allows a client to retrieve and update geospatial data encoded in Geography Markup Language (GML) from multiple Web Feature Services. More specifically, the OGC WFS [3] defines an HTTP-based interface for a data access service that enables features from multiple vector (feature) repositories to be queried and managed. The standard defines operations that enable clients to do the following:

- Discover which feature collections the service offers (GetCapabilities)
- Get a description of the properties of features (DescribeFeatureType)
- Query a collection for a subset of features that satisfy some filter expression (GetFeature)

- Lock a subset of features (LockFeature, GetFeatureWithLock)
- Execute transactions against feature collections (Transaction) that allow a client to create new features (Insert), modify existing features (Update), and delete feature (Delete)

The mandatory encoding for input and output is GML. However, the standard is extensible and allows for other feature encodings to be supported, such as GeoRSS* and KML[†].

5.5.4 *Open GIS Geography Markup Language (GML)*

The OpenGIS® GML Encoding Standard [4] defines a data encoding in XML for geographic data and a grammar to express models of such data using XML Schema. GML provides a means of encoding geographic information for both data transport and data storage, especially in a Web context. GML is extensible, supporting a wide variety of spatial tasks, from portrayal to analysis. It separates content from presentation (graphic or otherwise), and permits easy integration of spatial and non-spatial data. Clients and servers with interfaces that implement the OGC WFS interface read and write GML data. GML is also an ISO standard (ISO 19136:2007).

GML contains a rich set of primitives that are used to build application specific schemas or application languages. These primitives include the following:

- Feature
- Geometry
- Coordinate Reference System
- Time
- Dynamic feature
- Coverage (including geographic images)
- Unit of measure
- Map presentation styling rules

More recently, based on experience gained from using GML in the CAD-GIS domain, the members have been addressing requirements for encoding simple solid geometries, such as a general convex polyhedrons element with tetrahedron, pyramid, wedge, and hexahedron sub-types.

One of the powers of GML is that information communities, such as AEC, can define application schemas of GML that capture the information models required for interoperability in their respective communities. Application schemas are normally designed using ISO 19103 conformant UML, and then the GML application created

---

* GeoRSS is a de facto standard that defines how to encode location in RSS feeds.
[†] KML, originally known as the Keyhole Markup Language, is an OGC standard. KML was submitted by Google to become an international standard.

by following the rules given in Annex E of ISO DIS 19136.There are currently dozens of GML application schemas (for examples, go to http://www.ogcnetwork.net/node/210).

Germane to the interchange of urban model information in the CAD-GIS-BIM community is CityGML.

5.5.5 OpenGIS CityGML

CityGML [5] is a common information model for the representation of 3D urban objects. CityGML defines the classes and relations for the most relevant topographic objects in cities and regional models with respect to their geometry, topology, semantics, and visualization properties. Included are generalization hierarchies between thematic classes, aggregations, relations between objects, and spatial properties. The thematic information supported in CityGML goes beyond graphic exchange formats and allows the application to employ virtual 3D city models for sophisticated analysis tasks in different application domains like simulations, urban data mining, facility management, and thematic inquiries.

By way of clarification, "City" is broadly defined to comprise not just built structures, but also elevation, vegetation, water bodies, "city furniture," and more. Included are generalization hierarchies between thematic classes, aggregations, relations between objects, and spatial properties. CityGML is applicable for large areas and small regions and can represent the terrain and 3D objects in different levels of detail simultaneously. CityGML is structured such that either simple, single-scale models without topology and few semantics, or very complex multi-scale models with full topology and fine-grained semantic differences can be represented. CityGML enables loss-less information exchange between different GIS and users.

CityGML is implemented as a GML application schema. As an application schema, CityGML takes advantage of other open standards and its development has proceeded in careful cooperation with other groups. For example, graphical rendering of data encoded in CityGML can be accomplished using standardized computer graphics data formats like VRML,* GeoVRML, X3D,† or Universal 3D (U3D).

Work on CityGML began in 2002 and was initiated by the members of the Special Interest Group 3D (SIG 3D) of the Geodata Infrastructure North-Rhine Westphalia (GDI NRW) activity in Germany. The SIG 3D is an open group consisting of more than 110 companies, municipalities, and research institutions working on the development and commercial exploitation of interoperable 3D models and revisualization.

CityGML version 1.0 was approved as an OGC standard on August 22, 2008.

---

* Virtual Reality Modeling Language is an ISO standard: http://en.wikipedia.org/wiki/VRML.
† X3D is an ISO standard developed by the Web3d Consortium: http://www.web3d.org/x3d/.

5.5.6 OpenGIS KML 2.2 Encoding Standard

OpenGIS KML Encoding Standard 2.2 [6] is an XML grammar used to encode and transport representations of geographic data for display in an earth browser.* As such, KML is a language focused on geographic visualization, including annotation of maps and images. Geographic visualization includes not only the presentation of graphical data on the globe, but also the control of the user's navigation in the sense of where to go and where to look. KML uses a tag-based structure with nested elements and attributes and is based on the XML standard. From this perspective, KML is complementary to most of the key existing OGC standards including GML, WFS, and WMS. Currently, KML 2.2 utilizes geometry elements derived from GML 2.1.2. These elements include point, line string, linear ring, and polygon.

Google submitted KML (formerly Keyhole Markup Language) to the OGC to be evolved within the OGC consensus process with the goal of making KML Version 2.2 an adopted OGC implementation standard. Future versions may be harmonized with relevant OGC standards that comprise the OGC standards baseline. There are four objectives for this standards work:

- That there be one international standard language for expressing geographic annotation and visualization on existing or future Web-based online and mobile maps (2D) and earth browsers (3D)
- That KML be aligned with international best practices and standards, thereby enabling greater uptake and interoperability of earth browser implementations
- That the OGC and Google will work collaboratively to ensure that the KML implementer community is properly engaged in the process and that the KML community is kept informed of progress and issues
- That the OGC process will be used to ensure proper life-cycle management of the KML candidate standard, including such issues as backward compatibility

KML 2.2 was formally approved as an OGC standard on April 16, 2008.

5.6 OGC Interoperability Initiatives for 3DIM Domain Interoperability

The OGC Interoperability Program (http://www.opengeospatial.org/initiatives) provides an industry consensus, rapid engineering process to develop, test, demonstrate, and promote the use of interfaces and protocols that enable interoperable geoprocessing. The Interoperability Program organizes and manages Interoperability Initiatives that address the needs of industry and government sponsors. In addition

---

* Please read the preamble to the standard to better understand why Google submitted KML to the OGC to become an international standard.

to requirements, sponsors also provide financial support to cover architecture and administration fees as well as some indirect participant expenses, such as work that would not be done outside the test bed environment and travel. In a test bed, there are also participants. Participants are OGC members that provide scientists and engineers. These individuals collaborate with the sponsors to insure that geospatial interoperability requirements are addressed. The process typically involves the following:

- Defining use cases based on interoperability requirements
- Defining the test bed engineering architecture, including interfaces and data sources
- Defining which OGC, ISO, and other standards will be used and how they will be used
- Defining new interface and encoding standards as required
- Implementing the architecture
- Demonstrating the results of the work

Every test bed has a demonstration. In the demonstration, detailed scenarios are used to provide the framework within which the standards-based technology solutions defined by the participants show how the sponsor's requirements have been addressed.

There are different types of interoperability initiatives. However, for the OGC CAD-GIS-BIM interoperability testing activities, the test bed has been extremely useful. This is because the test beds are collaborative, applied research and development efforts to develop, architect, and test candidate standards addressing sponsor requirements. Further, the engineering cycle for a test bed is typically less than 6 months duration and the sponsors quickly receive return on their investment in the activity*.

The OGC has successfully completed five major test beds that have focused on OGC Web Services. A major interoperability focus area for the OGC Web Services-4 TestBed (www.opengeospatial.org/projects/initiatives/ows-4) was CAD-GIS-BIM integration in well-defined workflows.

5.6.1 OGC Web Services 4 (OWS-4) CAD-GIS-BIM (CGB)

The OWS-4 TestBed was an initiative of the OGC Interoperability Program to collaboratively extend and demonstrate the OGC standards baseline for geospatial interoperability as it relates to the built environment.

The activities of the CGB thread resulted in the development of several new types of components that demonstrate the integration of BIM standards with the OGC Web Services Architecture:

- Testing of a Transactional Web Feature Service (WFS-T for BIM) that serves features from BIM in both Industry Foundation Classes (IFC) and CityGML

---

* Sponsors have often received a 3-to-5 times return on their financial investment.

- New client capabilities for three-dimensional thematic viewing and analysis of building information in CityGML form direct from Web Feature Services
- New capabilities in BIM authoring clients that use CityGML from WFS and images from Web Map enabled services (WMS) to allow the development of BIM in geographic context

The test bed involved several interoperability experiments that stretched the capabilities of existing OGC services to support the new problems of serving CityGML using the OGC WFS standard. This test bed also resulted in fruitful discussions between the primary custodians of IFC and CityGML concerning the useful overlap of these two means of representing places.

To test the use of OGC standards and provide the technical foundation for the demonstration, the following use cases were defined:

1. Transactional Access to WFS-T for BIM: A space planner requests information about a building's space layout (schematic) from a WFS-enabled database. This information is returned as an IFC payload encoded as GML. The planner uses an editor client to create a new space scheme for this building. The new scheme is returned to the server.

2. BIM Authoring with Web Map Service Context Data from Open Web Services: The designer queries an OGC catalogue service for photography and CityGML City Model feature collections and IFC space information overlapping with the bounding box of his area of interest. The catalogue service returns the URL for a WFS, WMS, and query strings that will retrieve the appropriate contextual and building information. The BIM authoring client retrieves CityGML, WMS, and IFC information from various servers. This information is then integrated with detailed building model information, either stored locally or retrieved from a remote server. The designer uses this information to plan his building intervention. All of this information is integrated into design documents and building information models that are passed to builders, architects, and others involved in the AEC process.

3. Analysis of Building Information over Broad Geographic Area: A planner needs information on the building space attributes for all of the buildings in a broad area of interest. Through a query to an OGC Catalogue Service he discovers services with useful information and retrieves building space information for several buildings from WFS for BIM and other city features and terrain from a WFS containing CityGML City Models. In addition he retrieves aerial photography from a WMS. All of this information is combined, and several thematic displays and analytical reports are generated that help the planner make a decision.

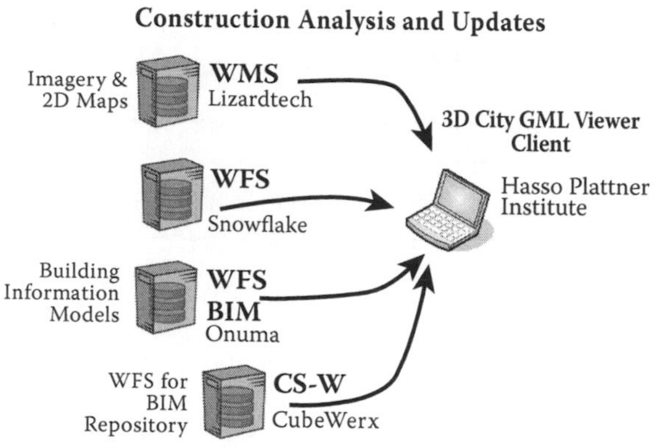

Construction Analysis and Updates

Figure 5.1 **Construction analysis and updates (Source: OGC).**

This test bed activity is described in detail in [7]. There is also an online video that can be downloaded and viewed (http://www.opengeospatial.org/pub/www/ows4/index.html). The OGC member participants in this test bed activity are listed in the footnote.* The list is provided to show that interoperability and standards development is truly a collaborative and consensus process.

5.6.1.1 How BIM Authoring with Context Data Using Web Services Was Accomplished

In terms of the BIM authoring and editing demonstration, the following was required. First, a client application was developed that allowed the operator to (1) find imagery and planimetric map data in their area of interest, (2) display data layers as required, (3) discover and access BIM structured building information for the structure of interest, (4) edit the content, and (5) post the results back into the source building structure (CAD) database.

For the demonstration, the imagery and 2D planimetric data were provided using several servers that were Web Map Service enabled. The CityModels were provided as GML via a WFS interface to multiple servers. BIM data was provided using another WFS enabled server. Figure 5.1 provides a high-level diagram of how the demonstration components work together.

---

* AEC3, AEC Infosystems, Autodesk, Bentley Systems, University of Bonn, Forschungszentrum Karlsruhe, General Services Administration, Harvard University Graduate School of Design, Hasso-Plattner Institute, lat/lon, LizardTech, OGC, Onuma Inc, National Institute of Standards and Technology, Parsons Brinckerhoff, Snowflake Software, Traverse Inc.

5.6.2 *Architecture, Engineering, Construction, Owner, and Operator (AECOO) Test Bed [8]*

In 2007, the OGC, potential test bed sponsors,* and the buildingSMART† alliance began discussing how the community could better create an environment in which key AEC workflow interoperability pain points could be defined in detail and then to actually proceed to test technology approaches that could solve these pain points. The discussions also touched on how to help orient the AEC standards community on the use of OGC test bed activities to accelerate the development, testing, and validation of new standards to address urgent interoperability problems. Such a test bed provides an opportunity to encourage multiple standards organizations to work across their mission lines to achieve results greater than what they can achieve by working in isolation. The OGC benefits from such a test bed by (1) being able to work in future joint initiatives to continue addressing the convergence issues, (2) bringing new interoperability requirements into the OGC process, and (3) stress testing existing OGC standards in new workflows.

Therefore, in February 2008 [9], the buildingSMART alliance and the OGC released a call for technology (RFT) in support of a new OGC test bed activity focused solely on three areas of interoperability activity of importance to the building industry:

- Decision support and general communications—connecting building models with business processes: What information needs to be exchanged between software systems that support formalized business processes, including design management, construction management, contract-required communication, and others; and how is this information integrated with the BIM and contract documents that are developed in parallel with these processes?
- Building Performance and Energy Analysis (BPEA): The test bed explores the interoperability between the expression of architectural design intent as captured by a BIM and the building's thermal loads. Also, the test bed explores interoperability between architectural models and annual energy performance with a focus on EnergyPlus and/or DOE2.
- Quantity take-off (QTO) for cost estimating: The role of the cost estimator is to facilitate the design process by systematic application of cost criteria so as to maintain a sensible and economic relationship between cost, quantity, utility, and appearance. Building modelers and cost estimators need to continually work together and share relevant information.

---

* The current AECOO test bed sponsors are: buildingSMART alliance, OGC, Associated General Contractors of America, American Institute of Architects, Burt Hill, Ellerbe Becket, Gilbane Development Corporation, HOK, Large Firm Round Table, NIST, Statsbygg (Norway), U.S. General Services Administration, Webcor Builders
† The buildingSMART alliance™ has been established to coordinate the profound constructive changes coming to the fragmented real property industry in North America. The organizations collective goal is open interoperability and full life-cycle implementation of building information models.

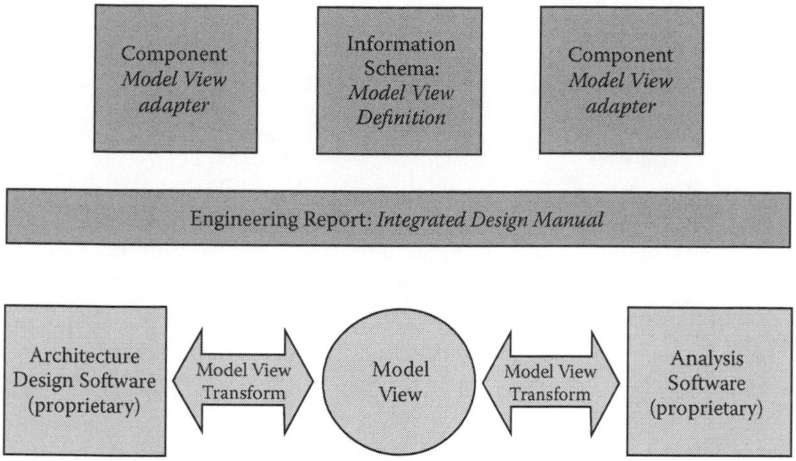

Figure 5.2   Engineering report: *Integrated Design Manual* (Source: OGC).

In terms of CAD-GIS-BIM Integration, agreement on information models and content are critical to data exchanges that are required during the completion of any workflow used in the building industry. This is why there is such a focus on information sharing in each of the three focus areas. Figure 5.2 abstracts the concept of information and model sharing between and among proprietary technologies.

Additionally, the test bed sponsors identified three problem areas related to how building models need to behave with cost estimating and energy programs once quantities have been input. The focus of this work effort is about three important items that cross the entire building life-cycle:

1. Building information platforms do not reliably pass to each other information useful for preliminary cost estimating during early phases of design.
2. Construction management software and building information platforms cannot share cost information interchangeably.
3. Building information platforms do not reliably interchange the information needed to conduct energy-related requirements scoping and analysis during early phases of design.

The following are the test bed objectives:

■ Demonstrate service* and information standards for model views and building information query and display.
■ Demonstrate data messaging using service and information standards for workflows.

---

* A service standard is a specific interface standard that provides for a given set of operations.

- Demonstrate data reusability and seamless machine interpretability.
- Incorporate transparency, accessibility, and usability of electronic information for exchange using access controls and data security provisions.
- Demonstrate using clients and servers together with tradecraft practices for information discovery, sharing, publishing and editing.
- Demonstrate collaborative tools cost estimating and energy analysis in a rich internet application setting that includes trades.
- Demonstrate methods for model query, repository, and management to include run time builds as well as published artifacts.
- Demonstrate quality assurance methods and procedures for data.
- Build and demonstrate reference test cases that enable software vendors and end users to know whether specific products can interoperate effectively.

This test bed is designed to have multiple components, each building on the other. The first phase of this test bed will be completed in the second quarter of 2009. The following deliverables will be made publicly available:

- Demonstration materials
- A number of engineering reports, such as "ifcXML Model View Definition for Quantity Takeoff," that document the test bed process, lessons learned, and possible changes to existing standards
- A number of proposals for changes to existing OGC or BIM standards
- Proposals for candidate OGC standards

These documents will be made available from multiple Web sites.

5.7 Summary and Conclusions

Solving the data sharing and workflow management issues related to the flow of information up and down the building life-cycle value chain is extremely hard. Only through collaboration between multiple information communities, such as AEC, GIS, remote sensing, surveying, and construction can we even begin to see tangible success in defining, deploying and using standards. The OGC membership is dedicated to working standards issues related to the geospatial aspects of the total built environment life-cycle. With billions of dollars in savings at stake, we have an imperative to define, test, and deploy standards that reduce costs and maximize the effectiveness of all professionals in the Architecture, Engineering, Construction, Owner, and Operator (AECCO) space and leverage the staggering investment in digital documents generated in support of the design, construction, and maintenance of our built infrastructure.

While there is still much work to be done in terms of realizing the vision of true CAD-GIS-BIM, the OGC and our partners have successfully opened cross-community communications and activities. This is a critical first step in truly solving and implementing true interoperability throughout the entire building life-cycle.

References

[1] Michael P. Gallaher, Alan C. O'Connor, John L. Dettbarn, Jr., and Linda T. Gilday, *Cost Analysis of Inadequate Interoperability in the U.S. Capital Facilities Industry*, National Institute of Standards and Technology, 2004.

[2] Jeff de la Beaujardiere (editor), *OpenGIS® Web Map Server Interface Standard Version 1.3*, OGC and ISO, 2006 (http://portal.opengeospatial.org/files/?artifact_id=14416).

[3] Peter Vretanos (editor), OpenGIS® *Web Feature Service Interface Standard Version 1.1*, OGC, 2005. (http://portal.opengeospatial.org/files/?artifact_id=8339).

[4] Clemens Portele (editor) (and Simon Cox, Paul Daisy, Ron Lake, Arliss Whiteside co-editors), OpenGIS® *Geography Markup Language Encoding StandardVersion 3.2.1*, OGC and ISO 2007. (http://portal.opengeospatial.org/files/?artifact_id=20509).

[5] Gerhard Gröger, Thomas H. Kolbe, and Angela Czerwinski (editors), *City Geography Markup Language—CityGML*, OGC 2007.

[6] Tim Wilson (editor), OpenGIS® KML Encoding Standard 2.2, OGC 2008.

[7] Paul Cote (editor), *OGC Web Services Architecture for CAD GIS and BIM*, OGC 2007. (http://portal.opengeospatial.org/files/?artifact_id=21622).

[8] Raj Singh, *Overview of the AECOO Test Bed*. OGC 2008. (http://portal.opengeospatial.org/files/?artifact_id=28895).

[9] OGC and BuildingSmart Staff, *AECOO Request for Technology*, OGC 2008. (http://www.opengeospatial.org/projects/initiatives/aecoo).

Chapter 6

CAD/GIS Integration Issues for Seamless Navigation between Indoor and Outdoor Environments

Mahsa Ghafourian and Hassan A. Karimi

Contents

6.1 Introduction

Navigation is defined as the task of tracking the movement of an object (e.g., vehicle) from one place to another [1]. With respect to land-based navigation, two types of navigation are discussed in the literature: indoor navigation and outdoor navigation. Indoor navigation systems assist users with navigation within buildings and outdoor navigation systems assist users with navigation in the outdoors.

Although there are currently a variety of technologies, techniques, and systems for navigation in outdoors or indoors, research that addresses navigation in both outdoors and indoors is limited. Consequently, users may need to carry different gadgets if their navigation needs include both outdoor and indoor environments. Such a lack of a navigation system that can assist users in navigating both outdoors and indoors is one of the several shortcomings of modern navigation technology. Other shortcomings include a lack of features to meet the navigation needs of physically, cognitively, and sensory impaired individuals, a lack of features to address navigation with respect to time of day (e.g., navigation in day vs. navigation at night), and limited usability for specific devices and navigation environments. In other words, today's navigation assistance systems (both outdoors and indoors) are one-size-fits-all and do not offer features that are universal. To address the shortcomings of current navigation systems, in this chapter we discuss the concept of universal navigation, called Universal NAVIgation Technology (UNAVIT), which we define as a navigation framework that provides navigation assistance anywhere, anytime, and for any users; a complete discussion of UNAVIT is beyond the scope of this book chapter and the readers interested in further readings on UNAVIT are referred to [2–4]. Furthermore, due to the pervasive nature of mobile devices and advances in key technologies such as geopositioning and wireless, such a universal navigation framework is anticipated to be available and accessible through smart phones.

Considering that development of universal navigation is a major undertaking where the integration of different technologies and databases is needed, in this chapter we focus on the anywhere feature of UNAVIT. For UNAVIT to facilitate anywhere navigation, it must provide seamless navigation between indoors and outdoors, which will be based on a holistic navigation approach as opposed to the fragmented approach taken by current navigation technology.

Supporting the anywhere feature needs understanding of both types of navigation environments, i.e., indoor and outdoor. There are similarities and differences between indoor and outdoor navigation. For example, both require similar functionalities such as finding points of interest (POIs), routes, and directions. On the other hand, they address navigation at different scales; for example, outdoor navigation is based on the scale of town, city, county, state, and country, whereas indoor navigation is confined within buildings. For this, geospatial information system (GIS) models and data are suitable for outdoor navigation, while computer-aided design (CAD) models and data (though geographic coordinate systems could also be applied), are suitable for indoor navigation.

Data for outdoor navigation is either road network or sidewalk network, while for indoor navigation it is floor plan or hallway network. Also, the geopositioning technologies for outdoors are typically GPS and dead reckoning (DR), or an integration of them, while for indoors they are typically radio-frequency identification (RFID) and WiFi. Indoor and outdoor navigation are also different with respect to modes of travel. In the outdoors, individuals drive, walk, or ride a bike/wheelchair, whereas indoors, users usually walk or ride a wheelchair. Furthermore, users' preferences are different for outdoors and indoors. While for outdoor navigation such routing criteria as shortest distance, shortest time, toll avoidance, least intersection, among others, are in demand (some affected by traffic, weather, accident, constructions, and other factors), for the indoors, routes' criteria are mostly limited to shortest distance and feasibility [5].

Clearly, one major consideration for addressing the issue of the anywhere feature, where transition between indoors and outdoors is accomplished in a seamless manner, is the integration of GIS and CAD. These are two incompatible platforms designed and developed over time for different and sometimes complementary purposes. The purpose of CAD creation is different from GIS [6]. GIS is a computer system with the aim of capturing, storing, retrieving, analyzing, and displaying geographically referenced information [7]. CAD system, on the other hand, is a computer system dealing with spatial data with the aim of automating drafting tasks [8]), where it can be used for drawing both two-dimensional and three-dimensional models [7]. There are differences between CAD and GIS functions. Some spatial analysis functions that are available in GIS, such as buffering and spatial query, are not available in CAD. Similarly, CAD systems support some operations such as spatial conflict detection with more details than GIS [6]).

In this chapter, Section 6.2 describes a scenario to better realize the needs of integrated navigation between indoors and outdoors. Section 6.3 provides background on indoor and outdoor navigations. In Section 6.4 the requirements of indoor and outdoor navigations are discussed. Section 6.5 focuses on issues related to CAD/GIS integration in indoor/outdoor navigation. In Section 6.6, two key algorithms for transition from indoor to outdoor and from outdoor to indoor are described. Section 6.7 provides a summary of the chapter.

6.2 Scenario

Peter, a new graduate student in the School of Information Sciences (SIS) at the University of Pittsburgh, lives in Columbus, Ohio. He has an appointment with his advisor to register for his courses at SIS and this is the first time he has traveled to Pittsburgh. Peter uses his smart phone equipped with UNAVIT, which provides navigation activities for all cities and for both indoor and outdoor environments. Below is his experience with UNAVIT. For his road trip, Peter requested the most scenic route between his current location (computed by GPS in his smart phone) and his new apartment in Pittsburgh. To ensure that he would arrive in time for his meeting with his advisor, in addition to a scenic route, he had included an approximate arrival time. UNAVIT provided a route based on both scenery and time criteria and Peter made it to his apartment in time. He then requested UNAVIT to find the shortest route from his apartment to the location of his advisor's office in the SIS building. He received instructions from UNAVIT on the route to reach the front of the SIS building. When he entered the building, UNAVIT directed him to the elevator and indicated the floor he needed to get off. Once out of the elevator on the floor where his advisor's office was, UNAVIT directed him to the office.

6.3 Background

Navigation anywhere, i.e., receiving navigation assistance both indoors and outdoors using the same navigation platform, is one major feature of universal navigation. Currently, there are research projects addressing the issues of indoor and outdoor navigation through the same platform. However, most of these research projects are focused on integration of geopositioning technologies, such as GPS and RFID, and require that the user make changes manually when moving from outdoor to indoor or vice versa.

Drishti [9] is an integrated navigation system that aims at assisting visually impaired individuals both in indoors and outdoors. Users can switch the system from outdoors to indoors with simple vocal commands. Users are provided with information on their current location and orientation, as well as direction to their desired destination (e.g., a sofa). Drishti provides visually impaired users with optimal routes, which include those with fewest hazards.

Harrison et al. [10] designed a seamless indoor/outdoor navigation system for ENABLED, a European framework for visually impaired individuals. The proposed system consists of two features: the hierarchical indoor/outdoor location detection system and the intelligent map. The first feature uses a hybrid indoor/outdoor geopositioning technique based on GPS and RFID. The intelligent map feature is a layered map that integrates with SVG/GML specifications.

NAVIO [11] is a pedestrian navigation system that can be applied both indoors and outdoors. NAVIO combines different positioning technologies such as GNSS, cellular phone positioning, and dead reckoning to assess the location and orientation of users. To provide seamless transition between indoor outdoor navigation, it uses a multisensor fusion model based on an extended Kalman filter.

Indoor/Outdoor pedestrian navigation [12] is a pedestrian positioning method aimed at merging DR, GPS, and RFID positioning. The experiment of the proposed research shows that GPS and active RFID tag systems transparently adjust estimation errors in DR.

Unlike other research projects, which are focused on addressing issues related to specific indoor/outdoor technologies or specific indoor/outdoor applications, in this chapter we focus on issues related to GIS and CAD integration upon user movement from outdoors to indoors or vice versa.

6.4 Indoor/Outdoor Navigation Requirements

Navigation systems capable of seamless navigation between outdoors and indoors must support features that are of importance for a variety of navigation purposes. A summary of outdoor and indoor navigation features is presented in Table 6.1.

Supporting these outdoor/indoor features in navigation systems requires certain adjustments in the new environment. These adjustments include several parameters, as summarized in Table 6.2.

To better understand the issues related to indoor/outdoor transitions, especially GIS and CAD, the following questions are analyzed:

- What buildings are contained within a given geographic area? GIS database contains address or coordinates of each building as a point, the footprint of each building, or both.
- Where are the entrance/exit doors of a given building? These could be coordinates of each entrance/exit door on the boundary of the footprint.
- What is the shortest route between two buildings? The route could be between the two points representing the buildings or between an entrance/exit door of one building and an entrance/exit door of another.
- In transitioning from outdoor to indoor, what data in GIS needs to be used and within what range? This could be map-matched data using a sidewalk network connected to entrance doors.
- In transitioning from indoor to outdoor, what data in CAD needs to be used and within what range? This could be map-matched data using the hallway network connected to exit doors.

6.4.1 Analysis

Geographic information system (GIS) is defined as a system for storing, retrieving, and analyzing geographical referenced data [13]. GIS is a tool for analyzing the outdoor navigation environment, primarily in two-dimension. CAD is a system for drawing and updating maps, [13] which is a tool for analyzing the indoor navigation environment, both in two-dimension and three-dimension.

To better understand the dimension issue as it relates to navigation in GIS and CAD, we present two simple scenarios. (1) the user needs navigation information to arrive at a destination (e.g., a restaurant) where the horizontal location (2D) of the user needs to be known and the altitude information is not crucial. (2) the user needs navigation information within a building, where location, both horizontal

Table 6.1 Outdoor/Indoor Navigation Features

Feature	Outdoor	Indoor
POIs	Different types Potentially many Often acquired through real-time geocoding	Few types Limited number No need to real-time geocoding
Routes	Distance and time are widely chosen criteria for computing optimal routes Real-time information such as traffic and weather affect the choice of routes Large gaps between personalized routes that are based on needs and preferences Distances between O-D could potentially be long Optimal routes affected by 2D/3D data Many route options between O and D	Distance may be the only criterion for most people No real-time information affects the choice of routes Individual needs and preferences for routes are not widely different Distances between O-D are short and limited to the size of the building Optimal routes based on 2D/3D data Limited route options between O and D
Navigation	Changes in mode of travel are possible Navigation impacted by factors such as weather, traffic, construction, and time. Accuracy within 10 meters	Change in mode of travel is not possible Navigation is not impacted by factors such as weather, traffic, and time. Accuracy within 3 meters

(x and y coordinates on a floor level) and vertical (z coordinate represents the floor level), needs to be known.

6.4.2 Data Model

For outdoor navigation, depending on the mode of travel (driving, walking, or biking), either a road network or sidewalk network is used. The user may change his/her location from road network to sidewalk network or vice versa, while changing mode of travel from driving to walking, from walking to riding, and so on. For

Table 6.2 Outdoor/Indoor Navigation Parameters

	Analysis	Data Model	Coordinate Systems	Scale	Geo-positioning	Modes of Transportation	Functions	User Preferences
Outdoor	GIS 2D/3D	Road Network, Sidewalk Network	Geographic Coordinates	Town, City, County, State, Country	GPS+DR	Driving, Walking, Riding Bikes, Riding Wheelchairs	Routes, Directions, POIs, LandMark	Shortest path, avoid toll, shortest time, most scenic
Indoor	CAD 2D/3D	Floor Plans, Hallway Network	Cartesian Coordinates	Building	RFID, WiFi	Walking, Riding Wheelchairs	Routes, Directions, POIs, LandMark	Feasible, comfortable, shortest path

indoor navigation, a hallway network or floor plan is used, and there is no change in mode of travel. As a result, navigation indoors is much more confined and simpler than navigation outdoors.

6.4.3 Coordinate Systems

Coordinate system is a referencing system used for measuring horizontal and vertical distances on a map [14]. The coordinate system used in outdoor navigation is primarily the geographic coordinate system based on latitude and longitude to assess the location of real-world features. Projected coordinate systems are also used in outdoor navigation. The coordinate system used in indoor navigation typically is based on a Cartesian coordinate system, though the geographic coordinate system could also be used. Cartesian coordinate system is a system that consists of two/three axes intersecting in the same origin, and that can be applied in two-dimensional or three-dimensional environments. The difference between the coordinate systems used in GIS and CAD is that geographic coordinate systems use longitude and latitude for representing the horizontal and vertical position with respect to the center of Earth, whereas Cartesian coordinate systems use a simpler set of axes for positioning with respect to an arbitrary origin [15].

6.4.4 Scale

Outdoor navigation may be at the scale of neighborhood, town, city, county, state, or country, whereas indoor navigation is only at the scale of buildings. Standardization of spatial data for navigation outdoors must take into account the differences at different scales, whereas standardization of spatial data for navigation indoors must be based on building features. An example of this standardization is the international standards offered by ISO/TC 211 for Geographic Information, which includes profile, spatial schema, temporal schema, spatial referring by coordinates, spatial referring by geographic identifiers, metadata, and so on [16].

6.4.5 Modes of Travel

In outdoor navigation, four different modes of travel are possible: driving, walking, riding a bike, or riding a wheelchair. In indoor navigation usually two modes of travel are possible: walking or riding a wheelchair. While it is possible a user's mode of travel in outdoor navigation might change (e.g., from driving to walking), there is no change in mode of travel in indoor navigation, i.e., individuals either walk or ride a wheelchair. Thus, with respect to functionality, outdoor navigation systems are more complex than indoor navigation systems.

6.4.6 Functions

Navigation both indoors and outdoors is based on fundamental functions such as finding POIs, computing optimal routes, and providing directions. Routing refers to the process of finding a desired route between a pair of origin and destination locations. Direction refers to the set of step-by-step instructions provided to the user in real-time to traverse a given route. POI refers to finding specific locations such as restaurants and theaters.

6.4.7 User Preferences

Users' navigation needs and preferences for routes in outdoors are constrained to networks used in outdoor navigation (road and sidewalk networks), different modes of travel, and various criteria affecting mobility (e.g., weather, traffic). Individuals could request shortest distance, fastest time, least intersections, and safe routes, among other criteria.

Unlike outdoor navigation, in indoor navigation, due to the simplicity of a building's structure and the limited navigation situations, user preferences are mostly confined to shortest or feasible routes. A feasible route refers to a route that can be taken by an individual given his or her special needs [5]. Issues such as traffic, safest route, and least intersection, do not exist in indoor navigation, thus, the solution space is very small.

6.5 GIS/CAD Integration

An ontology for outdoor/indoor navigation is shown in Figure 6.1. In this ontology, navigation consists of three concepts: indoor, outdoor, and function. Each concept of indoor and outdoor is consisted of geopositioning, mode of travel, scale, and model subconcepts. To be able to navigate seamlessly between indoors and outdoors, the integration between the concepts is depicted with dashed lines in the ontology.

In this ontology, seamless navigation between indoors and outdoors implies integration between CAD and GIS, such as networks and coordinate systems.

Upon the change of location from outdoor to indoor or vice versa, the coordinate system and the network must be adjusted to the new environment. For example, if a pedestrian enters a building the sidewalk network must be changed to a hallway network, and if a driver enters a building through a garage then the road network must be changed from road network to hallway network. Figure 6.2 depicts CAD/GIS integration for navigation.

To implement the aforementioned integration, in the next section the two required algorithms are discussed.

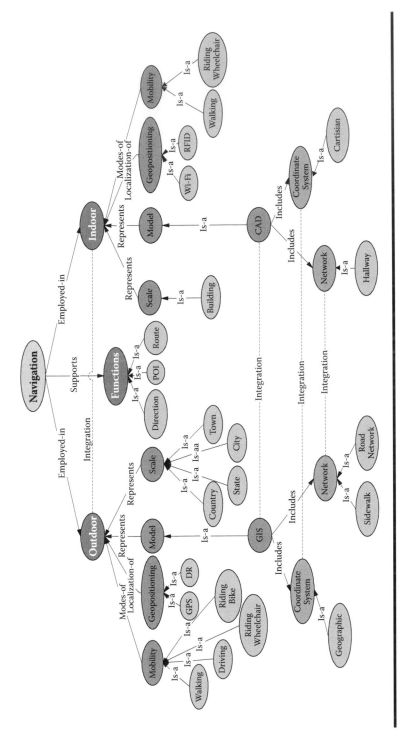

Figure 6.1   Ontology for integration between indoor and outdoor navigation.

Figure 6.2   GIS/CAD integration.

6.6 Algorithms

To develop a system that provides seamless navigation between indoors and outdoors, two key algorithms: Anywhere-OI and Anywhere-IO. The Anywhere-OI algorithm is used when the user is roving outdoors and approaches one of the entrance doors of a building. The Anywhere-IO algorithm is used when the user is in a building and approaches one of the exit doors. In both algorithms, all the mentioned parameters in Section 6.4 need to be adjusted to the new environment. In other words, upon the user's change of location from outdoor to indoor or vice versa, the underlying network (road, sidewalk, or hallway), database (GIS or CAD), geopositioning techniques, map-matching algorithms, and routing algorithms need to be adjusted to the new location.

In the Anywhere-OI algorithm (Figure 6.3), the user's mode of travel is taken into account. For a pedestrian, the user's position and trajectory with respect to buildings is constantly assessed (lines 1–5). As the user changes position on the trajectory, the upcoming building in the sidewalk segment is considered as the candidate building and the user's proximity to the building is calculated (lines 6–9). Once it is detected that the user is moving toward the entrance door of the candidate building and is about to enter the building, the transition from outdoor to indoor is made by making the appropriate adjustments (lines 10–13). For a driver, the algorithm checks to see if the driver is approaching the entrance of a garage (lines 16–18). Once it detects that the user has entered the building, the required adjustments will be made (lines 19–22). If the algorithm detects that the user has approached a garage door of a building but has not entered the garage, it assumes that the user has parked his or her car and started walking on the sidewalk. In this case, the algorithm checks to see if the user is about to enter a building (lines 23–26).

In the Anywhere-IO algorithm (Figure 6.4), the process is simpler, as there is only one choice for entering outdoors when the user's location is detected near an exit door of a building. Once it is determined that the user is approaching an exit door of the building, the algorithm checks the user's distance with regard to the

Input: user request for navigation information

// OBx,y is a point outside the entrance door of the candidate building
// IBx1,y1 is a point in a specific distance to the entrance door of the candidate building
// IBx2,y2 is a point in a specific distance to the entrance door, where IBx2,y2 < IBx1,y1 with respect to the entrance door
// EBx,y is a point represents the entrance door
// OGx1,y1 is a point in a specific distance to the entrance of a building's garage
// EGx,y is a point represents the entrance door
//   a threshold value

1.   POx,y ← Get user's outdoor position;
2.   If the user uses sidewalk network then
3.   Determine current sidewalk segment;
4.   Determine user's orientation;
5.   Retrieve candidate building from DB;
6.   If (|POx,y − OBx,y| <   ) then
7.   While (|POx,y − IBx1,y1| <   ) do
8.   While (|POx,y − IBx2,y2| <   ) do
9.   If (|POx,y = EBx,y) then
10.   Change to CAD/DB corresponding to the building;
11.   Change to hallwaynetwork of the building;
12.   Change map matching to pedestrian in hallway network;
13.   Change to routing algorithms for indoors;
14.   Else
15.   Anywhere-OI;

16.   If the user uses road ţ
17.   If (|POx,y − OGx1,y1| <   ) then
18.   If (|POx,y = EGx,y) then
19.   Change to CAD/DB corresponding to the building;
20.   Change to hallwaynetwork of the building;
21.   Change map matching to pedestrian in hallway network;
22.   Change to routing algorithms for indoors;
23.   Else
24.   Change to sidewalk network;
25.   Change map matching algorithms to pedestrian on hallway network;
26.   Anywhere-OI;
27.   Else
28.   Wait for receiving POx,y of the user.

Figure 6.3   Anywhere-OI algorithm.

exit door (lines 1–3), and if it determines that the user is about to exit the building, the transition from indoor to outdoor will be made by making the appropriate adjustments (lines 4–6). Furthermore, if a user enters the garage of the building to drive outside, the system can recognize and wait to acquire the user's next position (lines 8–10). If the next position indicates that the user is outside of the building,

```
Input: user request for navigation information

// EBx,y is a point represents the building exit door
// OBx,y is a point represents outside of the building
// Gx,y is a point represents the building's garage

// δ a threshold value

1.        PIx,y ← Get a user's indoor position;
2.      If(|PIx,y - EBx,y| < δ) then
3.          If(|PIx,y - OBx,y| < δ) then
4.              Change to GIS/DB;
5.              Change to sidewalk network;
6.              Change to map matching to pedestrian on sidewalk network;
7.      Else
8.              If(|PIx,y - Gx,y| < δ) then
9.                  If PIx,y is not been receiving then
10.                     Wait for receiving POx,y of the user.
11.                 If POx,y is acquired then
12.                     Change to GIS/DB;
13.                     Change to road network;
14.                     Change to map matching to driver on road network;
15.                 If PIx,y is acquired then
16.                     Anywhere-IO;
17.         Else
18.                 Anywhere-IO;
```

Figure 6.4   Anywhere-IO algorithm.

then the required adjustments will be made (lines 11–14). Otherwise, the algorithm starts all over again (line 16).

6.7 Summary

In this chapter we discussed issues pertaining to seamless transition between indoor and outdoor navigation. We presented features relevant to navigation and analyzed specific parameters that systems supporting indoor and outdoor navigation must take into consideration. To better understand the issues of integrating CAD and GIS, an ontology where integration of concepts and relationships were highlighted was presented. In this ontology, certain CAD and GIS parameters, including networks and coordinate systems, need to be integrated. We also discussed two algorithms, Anywhere-OI and Anywhere-IO that provide the overall logic of moving from outdoor to indoor and indoor to outdoor, respectively, and make the appropriate adjustments on the parameters required for the transition.

References

1. Bowditch, N., and Bowditch, J. (1880). *The American Practical Navigator*, Reprint Services Corp.
2. Ghafourian, M. and Karimi, H.A. (2009). Universal Navigation Concept and Algorithms. In *World Congress on Computer Science and Information Engineering (CSIE 2009)*, held in Los Angeles/Anaheim, CA. CD-ROM available from IEEE.
3. Ghafourian, M., Karimi, H.A., and Roosmalen, L. V. (2009). Universal Navigation through Social Networking. In *Proceedings of The 13th International Conference on Human-Computer Interaction (HCI 2009)* in San Diego, CA. Berlin: Springer-Verlag.
4. Karimi, H.A. and Ghafourian, M. (2009). Universal Navigation. *GIM International*, May 2009, 17–19.
5. Dudas, P., Ghafourian, M., and Karimi, H.A. (2009). ONALIN: Ontology and Algorithm for Indoor Routing. In *Proceedings of The 10th International Workshop on Indoor Navigation Awareness*. Taipei, Taiwan.
6. CAD to GIS: A Step-by-Step Guide to Converting .dwg Cad Files to Gis Shapefiles. (2005). National Park Service (NPS) [cited 14 January 2009]. Available from http://mms.nps.gov/gis/applications/documents/cadgis_rev1d.pdf.
7. Howell, I. and Batcheler, B. (2005). Building Information Modeling Two Years Later: Huge Potential, Some Success and Several Limitations. *Laiserin Letter* 24. Available from http://www.laiserin.com/features/bim/newforma_bim.pdf. Accessed Oct. 14, 2009.
8. Cowen, D. (November 1988). GIS Versus CAD Versus Dbms: What Are the Differences? *Photogrammetric Engineering and Remote Sensing* 54(11): 1551–1555.
9. Ran, L., Helal, S., and Moore, S. (2004). Drishti: An Integrated Indoor/Outdoor Blind Navigation System and Service. In Proceedings of the Second Annual Conference on Pervasive Computing and Communications, 14-17 March, in Orlando, FL. *PerCom '04*, 23–30.
10. Harrison, B., Wu, H., Marshall, A., and Yu, W. (2004). The Enabled Indoor/Outdoor Navigation Systems for the Blind and Visually Impaired. In *IEEE ICT,* Brazil.
11. Retscher, G., and Thienelt, M. (2004). Navio-a Navigation and Guidance Service for Pedestrians. *Journal of Global Positioning Systems*, 3(1-2): 208–217.
12. Kourogi, M., Sakata, N., Okuma, T., and Kurata, T. (2006). Indoor/Outdoor Pedestrian Navigation with an Embedded GPS/RFID/Self-Contained Sensor System. In *Proceedings of the 16th International Conference on Artificial Reality and Telexistence*. Hangzhou, China. Springer, 1310–1321
13. Curry, S. (May 2003). Autodesk Infrastructure Solutions. CAD and GIS: Critical Tools. Critical Links, white paper.
14. Yeung, A. K. W., and Lo, C. P. (2007). *Concepts and Techniques of Geographic Information Systems*. 2nd ed: Pearson Prentice Hall.
15. Review B: Coordinate Systems. (2002). Massachusetts Institute of Technology, Department of Physics.

Chapter 7

Semantics in CAD/GIS Integration

Michael J. Casey and Sriharsha Vankadara

Contents

7.1 Introduction

CAD and GIS have evolved on similar paths for decades. Both have foundations in early computer graphics, both have evolved to utilize modern software architectures, and both have established user bases, albeit in different domains. CAD is the predominant platform for architecture, engineering, and construction (AEC), while GIS is widely used for municipal planning and natural resource management. While CAD and GIS are closely related technologies, they remain largely distinct, as has been discussed throughout this book. CAD applications are generally focused on design, are applicable to the scale of buildings or infrastructure works, and often lack a robust attribute model. GIS applications are focused on spatial analysis, defined at geographic scales (from regional to global), and rely heavily on a database attribute model. Despite these differences, the need for *integrated* planning, design, and management of natural resources and infrastructure assets is increasing, and so is the need for integration between CAD and GIS.

The term *integration* has been used throughout this book as the ability for CAD and GIS applications to combine for a common purpose. A related term is *interchange* which focuses on data exchange only between systems. Still another related term is *interoperability* that carries a slightly different meaning: to allow the exchange of data *and* processing capabilities while allowing the participating systems to remain distinct. This separation is advantageous in that it permits the highly specialized and powerful functionality of each system to remain available (and presumably optimized for each system), while exposing the more mainstream functionality for *interoperation*. The spirit of this separation with the opportunity for integration is in keeping with the overall theme of the chapter, which is the treatment of *semantic interoperability*, to be defined shortly. The goal of this chapter is to further elaborate on the challenges and opportunities with CAD and GIS integration. Specifically, the unique issues of semantic integration are addressed. Semantic integration is the process of enabling interchange and interoperability between systems without predetermined interfaces. It is the process of encoding domain knowledge about an object or a phenomenon in such a way that software can discover facts about that object without prior agreement about the *meaning* of said facts.

CAD and GIS have somewhat different approaches for handling semantic properties. Table 7.1 summarizes these differences with respect to their respective treatment of geometry, data, and metadata. Generally speaking, modern CAD software performs exceptionally well with vector features organized into layers and blocks. The extension of vector features to include attribute data and metadata, however, is somewhat limited. This is not a limitation of any particular software, but rather

Table 7.1   Semantic Comparison between GIS and CAD Platforms

Platform	Class	Interfaces	Semantic Interchange Potential
CAD	Geometry	Vectors accessible by layer or sheet	Limited, elemental geometry
	Data	Attributes can be tagged to certain features, no attribute/ RDBMS model	High, if entities are tagged at functional level of abstraction
	Metadata	Limited to project or file	Limited
GIS	Geometry	Vector or raster accessible by file or geodatabase	Moderate, geometry is topologically integrated, linked through abstract class relationships
	Data	Attribute and spatial query possible	High, attribute model is accessible for semantic discovery
	Metadata	Desctriptive text, XML, stardards-based	High, data is coded with source and parameter definitions

an artifact of the CAD platform. By contrast, GIS enables a richer data model that permits extension of features to include data and metadata descriptions. The semantic interchange potential of these two related but distinct platforms serves as the backdrop for the rest of this chapter.

The pace of the development of methodologies for semantic integration is currently very rapid. In the United States, and through international standardization organizations, e.g., the World Wide Web Consortium (W3C), International Alliance for Interoperability (IAI), and Construction Specifications Institute (CSI), a lot of work is currently taking place to define the standard taxonomies, languages, and services to allow flexible definition of domain knowledge in the form of *ontologies*. Ontologies are formal definitions of domain knowledge expressed in software-readable languages that define the objects (representations) in a domain and how they relate to one another (relationships). Enabling technologies such as the semantic Web,[1] which is itself based on technologies such as XML and Resource Description Framework (RDF),[2] has provided universal accessibility to ontologies with foundations for software services capable of *reasoning* and making *inferences* based on domain knowledge in the form of ontologies.

The development of standard taxonomies and ontologies for CAD and GIS is still largely separate, but evolving. As will be discussed in more detail in the next section, standard representations and relationships for CAD objects, or product models, have evolved over many years and are now widely used in design, fabrication, and manufacturing.[3-4] For GIS, domain object models,[6] languages for open data interchange,[7] and ontologies are emerging.[2] The capability to combine the separate ontologies and languages from CAD and GIS for semantic interoperability is still limited, although some work has been done to meet this need.[9]

One technology that holds considerable promise for bridging the semantic differences between CAD and GIS is Building Information Modeling (BIM). BIM is a modeling technology that combines the design and visualization capabilities of CAD with the rich parametric object and attribute modeling of GIS. As its name implies, BIM is geared primarily toward building construction, although broader adoption by so called "horizontal" industries such as transportation and infrastructure utilities is occurring.[10] BIM technology is applicable to the entire lifecycle of buildings from initial feasibility planning, through design, construction, commissioning, and asset management.[11] The result is a rich, object-relational data model that facilitates interdisciplinary building design, construction, and operations.

BIM is premised on object intelligence. Geometric features are represented as complex, real-world objects with structure, behavior, and constraints relative to other objects. This differs substantially from the traditional CAD approach where objects are typically represented with simple geometries, line colors or weights, and across multiple layers. 2D CAD (in many cases) expresses design intent implicitly rather than explicitly. In other words, only details for typical sections are given, notes are provided for some parts of the design (but not all), and the overall design must be "synthesized" from a series of plan, section, and elevation sheets. BIM enables a complete 3D design representation. Plans and drawings are an arbitrary by-product of the model, not the focus of the design. With BIM, interdependent building systems (e.g., structural and mechanical) can be visualized, but most importantly designed in an integrated manner. Figure 7.1 depicts a typical 3D rendering of a BIM model developed by PSA-Dewberry for a federal courthouse in Rockford, Illinois. Using layered 3D visualization, the HVAC mechanicals within the building can be seen, and conflicts with the building's structural skeleton can be immediately detected and corrected.

BIM technology borrows heavily from the spatial indexing and attribute models (engines) that have been part of GIS for years. Specifically, BIM utilizes *topological integration*. Topological integration is the recognition of distinct vector features that share geometry. Rather than two adjacent polygons having a double boundary between them, topological integration allows the boundary to be shared. In BIM, topologic integration is implemented at a higher level of abstraction. Adjacent objects may share geometry such that constituent parts of, for example, a wall system, such as a window,

Figure 7.1 BIM model of the Rockford, Illinois Federal Courthouse. Image courtesy of PSA-Dewberry.

door, or access panel "understand" that they are related to one another by the same wall interface. Similarly, the parametric behavior of objects mirror those used in more recently developed geospatial object models. For example, models of a water distribution system that incorporate constraints that prevent certain pipe diameters or materials from being connected to one another are similar to behavioral rules in BIM that prevent certain building materials to be combined with one another. Such intelligence facilitates automated error detection and code checking, e.g., International Building Code (IBC) or National Fire Protection Association (NFPA) fire safety codes.

By serving as a kind of *hybrid* between CAD and GIS, BIM offers the opportunity to effectively model something that has been lacking for integrated building planning and design: the transition from *in-building* conditions and analysis (usually the realm of CAD) to the *out-of-building*, or site conditions (usually the realm of GIS). With BIM, it is conceivable that one can interoperate between the objects and behavior of the smaller scale in-building systems and the larger scale out-of-building ones in a seamless manner. There are significant challenges to this vision however. Specifically, the ability to effectively model the semantics, or meaning, across these domains and scales is largely undefined at this time.

This chapter will address the issues in semantic integration between CAD and GIS specifically in the context of BIM. Section 7.2 presents state of the practice perspectives on integration between CAD and GIS as well as a discussion of syntactic integration, and a more in-depth treatment of the issues with semantic integration. Section 7.3 describes some use cases for semantic CAD-GIS integration that can benefit from a BIM-based approach. Section 7.4 provides further background on BIM, including standardization efforts within and across several industries and some current software tools. Section 7.5 discusses a case study on the application of semantic modeling to facilitate the integration of CAD and GIS, through BIM, to the modeling of evacuation planning for campus environments. Details on the geospatial data and ontological support needed to provide comprehensive evacuation planning from inside offices and buildings to shelters (areas of refuge) will be discussed. Section 7.6 provides a chapter summary that recaps the current opportunities and challenges in CAD/GIS integration with respect to semantic modeling and the strengths that BIM can provide in bridging these platforms. Finally, Section 7.7 presents some conclusions about the enduring role of semantics in CAD and GIS integration.

7.2 Perspectives on Interoperability and Integration

The desire and capability to integrate disparate information systems is nothing new. Commercial Relational Database Management Systems (RDBMSs) and desktop software have a long history of success and failures in interoperability. GIS and CAD share something at their foundation that makes the potential for integration and interoperability seem greater. They share, at least nominally, the same 2D vector and layered graphics environment. They differ, drastically in some cases, other aspects including reliance on geo-referencing, query capabilities, and surface model-ing. These two platforms have come closer together recently, especially in terms of respective support in commercial applications. However, there remain many differ-ences that equate to obstacles for interoperability that are presented here. This sec-tion discusses first the syntactic approaches for integration, but primarily semantic approaches, and finally a discussion of open formats and domain knowledge repre-sentations appropriate for enhanced CAD and GIS integration.

7.2.1 Syntactic Interoperability

Syntactic interoperability is defined as the ability of a given software program or ser-vice to directly access data and methods from a different program or service for read-ing, writing, or both. Basic interoperability usually involves data read operations, while complex interoperability requires both read and write. The prerequisite for syntactic interoperability is an agreed upon file or service specification. This can be accomplished in two ways. First, the different producers of the formats may publish formal specifications that describe their respective data structures and how they may be implemented. Or, producers may elect to use an *independent* format specification that is vendor neutral.

Let us first consider the vendor published format specification. The syntactic interoperability provided through vendor-published format specifications promul-gates two important shortcomings:

1. The published format specification may hide some complex aspects of the data format or associate the most advanced features and capabilities of a ven-dor's product to only a closed or unpublished format. The motivation for both of these actions is usually to maintain some proprietary control or to protect a perceived competitive advantage.
2. Syntactic interoperability is based on the assumption that disparate systems must identify objects in the same way. A vector line object from one vendor may be specified as a connection between two Cartesian coordinate points, with one point designated as the origin and the other the destination. Another vendor may define the line object as having an origin point in a polar coor-dinate system with only a length and angle of elevation, φ. Interoperability requires the respective software programs or services to recognize the *common meaning* of the data structure, irrespective of its implementation.

A very well known example of an interoperable format is the AutoCAD Drawing Interchange Format or DXF[12]. DXF was first introduced in the early 1980s as an open (published) format specification for interchange of drawing data between CAD applications. Formats for both ASCII and binary are available, with support ranging from very early to very recent (i.e., AutoCAD 2010) versions of AutoCAD, as well as most other CAD packages. The format specification published by

Autodesk includes structures for BLOCKS, OBJECTS, ENTITIES, and others, which themselves include elementary geometries, and layers. DXF is by design an interchange format, not a native one, and is as such not optimized for primary use. For this purpose, drawings are preserved in the unpublished, proprietary DWG format. DWG can generally be shared with other applications through DXF conversion, although some loss of information may occur.

An example from the GIS industry is the ESRI shapefile format. The shapefile specification[13] describes the multi-file vector GIS format popular for dissemination of geospatial data in the public domain. It includes a detailed format description of basic feature geometries such as points, polylines, and polygons as well as specifications for the dBASE database format on which the shapefile attribute model is based. The published format specification includes guidance for exchanging geospatial data among the various ESRI products, as well as for direct access through low-level programming interfaces, or application programming interfaces (APIs). The modern ESRI equivalent to the shapefile is the Geodatabase. Direct data interchange between Geodatabases and other software is possible using the Geodatabase XML format schema, but the shapefile format has become a more common format used for interchange.

Through the years, several "data converters" have been developed which allow translation of one format into another based on these published format specifications. Further, most commercial CAD and GIS packages have provided software extensions to allow direct access to other formats (e.g., ESRI ArcGIS Data Interoperability).[13] The use of data converters for direct access to geospatial data from within CAD, and CAD data from within GIS, is almost always read only, and is even then subject to limitations. For example, the GIS typically expects all layers to be drawn in a predefined geographic coordinate system, not a "page space" as is common with CAD drawings. Similarly, CAD applications cannot usually deal with scale-dependent rendering which limits the level of detail in a map view depending on the view scale.

Returning to the second method of providing an agreed-upon format specification, let us now consider the independent format case. It is important to realize that the published format specifications provided by a specific vendor are not technically open. The controlling vendor may elect to drop or alter their compatibility with future versions of their proprietary formats at anytime. A movement to provide open standards for CAD and GIS data interchange is now mature with numerous open file and service specifications available. The U.S. National CAD Standard [15] and the Open Geospatial Consortium[7,16] both publish open standards and formats for file-based and service-based interchange. Open standards based on languages such as Geography Markup Language (GML)[7] have produced numerous technology demonstrations and open applications that alleviate some proprietary obstacles to data interchange.

One challenge to syntactic interoperability is the advent of rich data models that incorporate increasing levels of domain knowledge. Simple feature geometries can usually be translated or interchanged, but object classes with complex structure and behavior pose a more difficult problem. For example, a polyline representing the alignment of a pipe in geographic space is easily exchanged between a CAD and GIS package. Even data about a pipe with attributes such as material type, installation date, service history, etc. is straightforward to exchange between CAD and GIS, provided the data types are elemental (i.e., strings, integers). Complex behaviors, however, such as linear referencing for service laterals, and pipe material and size combinations are much more esoteric and are often lost in translation or interoperation efforts. Higher-level representations of objects as well as schemas and domain-specific knowledge representations are needed to facilitate interoperability with more richly defined objects.

Product data models such as the Industry Foundation Classes (IFC) and IFC/ifcXML[3] attempt to relate the specification hierarchies (e.g., CSI UniFormat™ and Uniclass™) used in the Architecture, Engineering, and Construction (AEC) design process to the hierarchies of AEC-related objects, their embedded intelligence, and their exposed interfaces. IFC is a neutral and open vendor standard for data representation and also a file format accessible by many CAD (and BIM) software applications. IFCs define architectural and construction CAD graphic data as 3D real world objects so that CAD users can easily transfer designs between different products belonging to a wide variety of vendors. IFCs have been developed based on 3D object-based CAD, which is currently the emerging CAD standard in the industry. Since IFCs describe buildings in the form of 3D objects, they are not usually applicable to the older 2D line based design methods. IFCs provide definitions for each object that exists in the building industry, and a text-based (ASCII) structure for storing object definitions. In this way, IFCs amount to an open format specification that goes beyond the simple geometry and attribute interchange possible with DXF and shapefiles. IFCs provide a richer object library that has been developed by industry with international consensus.

The IFCs and related efforts have evolved over a long period of time and are constantly changing. While it is a good thing that standards for product model exchange are keeping up with changes in the industry, BIM software vendors interested in supporting them directly in their software have been frustrated with increasing complexity, lack of backwards compatibility, and lack of extensibility.[11] Peachavanish et al.[9] discuss that support for semantics to manage the vendor's concerns is possible within the IFC framework, but that it has been intentionally omitted for one main reason: agreement about the semantic nature of all objects is not possible across all applications. For example, the commercial cost estimating software Timberline will read IFC data, but requires manual mapping of the IFC term definitions to the ones favored by Timberline.[9] The fundamental limitation listed as the second syntactic assumption above (i.e., the necessity of agreement) remains for both low level interoperability based on simple geometric features and attributes to high level interoperability based on domain specific product models.

The implementation of a published format standard can differ, but the *meaning* or *semantics* must be the same for interoperability to be possible.

7.2.2 Semantic Interoperability

Semantic interoperability is defined as the interaction of two or more systems sharing common data or methods *without* a predefined or agreed upon interface. By explicitly embedding the meaning or *semantics* with the data element to be exchanged, it allows direct access to an object's metadata and a flexible and extensible means to manipulate that object. The mechanics by which this occurs cannot be explained without a brief discussion of knowledge representation, metadata modeling, formal semantic languages, and logical reasoning.

The encoding of semantics about an object or phenomenon requires a means for knowledge representation. This is achieved by the creation of a domain ontology, which was introduced in Section 7.1. A domain ontology begins with a grammar or taxonomy of the important objects and classes within that domain. In this way it is like a schema for databases with a defined column structure and specified data types, or an XML schema with a tree structure and attributes for each node. An ontology goes beyond the concept of a schema in its ability to express relationships arbitrarily. Unlike schemas, which are constrained by elemental relationships like cardinality or set membership, ontologies allow abstract concepts to be defined and represented.

Casey and Austin[16] describe the development of an ontology to support spatial navigation. They present the scenario of a tourist trying to navigate through New York City's Central Park using a handheld device capable of processing semantic queries. The device is able to generate semantic data in response to the tourist's query by dynamically assembling an instance of several interconnected ontologies related to geographic location, availability of a GIS database, and applicable maintenance closures which may affect routing. The resulting ontology instance is a directed graph, which can be analyzed through logical reasoning to infer facts such as whether a particular path is open or closed. In other words, the ontology can be used to represent a complete route based on inferred information, rather than from a pre-defined geospatial data source.

It should be observed that ontologies are by their very nature open. If deployed on the WWW, they provide universal accessibility. When differing ontologies of a same or similar object are created, they may be reconciled by: 1) crawling and indexing of all known ontologies about or related to the object; and 2) invoking inferencing and reasoning processes, to be discussed below.

The language of ontologies is written in low-level logical expressions, commonly used for representing metadata. One example is the Resource Description Framework (RDF), which will be covered in detail later. This differs somewhat from the approaches used in representing GIS metadata. In GIS, metadata for a land use layer may indicate the data source, distributor, spatial reference, accuracy, etc. CAD metadata tends to be document or project specific. For ontology modeling, the metadata can capture more dimensions. For example, the land use layer, rather than having explicit land use code numbers that correspond to some classification scheme stored offline, can store subclass (container) relationships, parent-child relationships, sequences, and *not a member of* relationships to indicate direct relationships between the classes.

The expressivity of metadata languages like RDF turns out to be somewhat limited when more complicated relationships need to be formed. For example, when trying to assert that an object "is-a" type of something, it is possible that multiple "is-a" representations may exist, leading to a certain ambiguity. Formal semantic languages like the Ontology Web Language (OWL) introduce formal description logics that minimize these ambiguities.

Finally, the chance to perform logical reasoning and inference with an instance of an ontology represents the ultimate goal of the approach. Logical reasoning and inference are in fact the tools necessary to validate the entire process. They allow the assertions expressed in the low-level ontological knowledge representations to be proven or disproven, thereby generating facts about the domain that were not previously expressed or encoded.

The nature of distributed systems is predicated on legacy technologies, syntactic, and semantic differences. Semantic interoperability is the means by which separate and distinct information systems may exchange data, processing methods, and output without pre-defined interfaces. Semantic interoperability is self-reliant and allows layers of middleware to discover the respective systems interfaces.

Semantic interoperability is not a panacea. It still requires that domain knowledge be encoded in an efficient way that balances expressivity (completeness) with computational tractability. The promise of the technology is that the ontological descriptions of AEC business processes are finite, and that eventually, sufficient domain knowledge will exist to allow automated reasoning and inference with backward compatibility and support for extensibility in newer ontology versions. This is still subject to some human oversight and validation, but the point is to automate not eliminate conflation issues.

7.2.3 Semantic Web Languages and Services

The term *Semantic Web* was coined in 2001 to describe a layer of WWW data and services that provides machine readable markup, not for document formatting like Hyper-Text Markup Language (HTML) or cascading style sheets (CSS), but for machine readable ontological metadata descriptions.[1] A range of semantic Web languages for encoding metadata about Web services and their available interfaces have evolved since that time. The Semantic Web is designed to be exposed for reasoning agents that can interpret ontological definitions and make inferences—a powerful and still not fully realized dream of what the Web can be. The previous section presented a high-level overview of semantic interoperability beginning with ontologies that should read like a progression. In fact, Semantic Web languages are

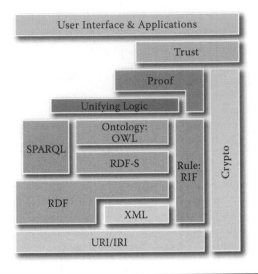

Figure 7.2   The Sematic Web layer cake (Copyright © 2007 World Wide Web consortium, [Massachusetts Institute of Technology, European Research Consortium for Informatics and Mathematics, Keio University] all right reserved. http://www. w3.org/2007/Talks/0130-sb-W3CTechSemWeb).

designed in this manner and build-upon the functionality of underlying layers as the well-known semantic Web "layer cake" shown in Figure 7.2 depicts.[1] Low-level interchange languages such as XML (extensible Markup Language) act as the foundation for the more expressive languages described below.

Resource Description Framework (RDF) is a metadata modeling language that is used as the basic building block for ontologies on the Web. RDF is based on simple expressions of subject-verb-object statements, called *triples*. Each resource in an expression is referenced as a universal resource indicator (URI) on the WWW. Arbitrary expressions can be created to make a statement about an object. Collections of these statements, which can be built up to whatever level of detail is desired, form instances of an ontology. Figure 7.3 shows a schematic representation of an RDF triple. An arbitrary object, in this case a geospatial feature ("point") is related to another arbitrary object ("polygon") using the relationship "Inside." Each part of the triple is referenced as a URI and is presumed to be universally accessible. RDF (and the related RDF schema) is usually written (serialized) in XML format. The directed graph structures of RDF are thus re-factored into XML trees that are easily manipulated by a wide variety of XML parsers and tools.

OWL is built on top of RDF to provide expressivity beyond the simple subject-predicate-object triples. OWL and its predecessor DAML+OIL (DARPA Agent Markup Language + Ontology Inference Layer) introduce logical expressions and operators for processing the statements expressed in RDF ontologies. OWL is

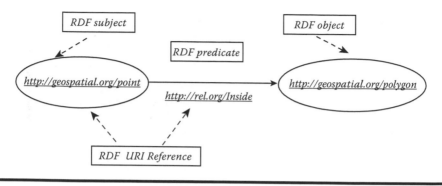

Figure 7.3   Example of an RDF triple.

divided into three sub-languages: OWL Lite, OWL DL, and OWL Full. OWL Lite is used for representing simple classification hierarchies with constraints. OWL DL introduces description logics for enhanced expressivity but also with computational completeness. OWL Full is reserved for maximum expressiveness with no guarantee for computational tractability. Table 7.2 summarizes some key language syntax and properties for each of the three languages.

OWL-S (Ontology Web Language–Services) is a markup language for the specification of semantic Web services. OWL-S brings the same arbitrary and flexible definition capabilities found in OWL to Web services. It provides three important features:

Table 7.2   Summary of OWL Sub-languages and Typical Language Constructs for Each

Sub-language	Purpose	Expressivity	Complexity	Application
OWL Lite	Hierarchical classification	Low	Low	Thesauri and Taxonomies
OWL DL	Encoding of description logics	High	Medium	Logical reasoning with guaranteed computational completeness
OWL Full	Maximum syntactic freedom	Very High	Very High	Formal logical reasoning without guaranteed computational completeness

1. **Automatic Web service discovery.** Automatic Web service discovery is the process of locating Web services that can provide a particular service or processing capability. The approach offers the user the flexibility of specifying the general parameters of the service or end result required, without specifying the mechanics of discovery. A semantic service or search agent can be tasked with discovering and brokering the interaction according to the parameters of the original request. The current syntactic approach requires a format or schema specification to be predefined. With OWL-S, the information necessary for discovery could be specified as semantic markup at the service Web sites, and an ontology-enabled search engine could be used to locate the services automatically.
2. **Automatic Web service invocation.** Automatic Web service invocation involves the invocation of a Web service given only a basic description of that service, as opposed to when the agent has been hard coded to call a particular service based on a set service specification.
3. **Automatic Web service composition and interoperation.** Automatic Web service composition involves the automatic identification, creation, and interoperation of Web services to perform a complex task, given only an abstract description of the desired objective.

These languages and techniques have been specialized for different domains. The next section presents the current tools and techniques for harnessing semantic Web technologies for geospatial modeling.

7.2.4 The Geospatial Semantic Web

Efforts to develop the Geospatial Semantic Web have evolved in parallel with the semantic Web languages and services. Efforts by OGC, as well as by academic consortia in the area of GIScience, have produced numerous testbeds and technology demonstrations that illustrate how basic semantic reasoning is possible given geospatial data with ontological definitions of domain knowledge. Instances of disparate domain knowledge can be combined in unexpected ways to produce new approaches for supporting decisions and inferences.[17]

7.2.5 Criticism of Semantic Interoperability

One could argue that the requirement that a domain ontology be shared to facilitate interoperability at the semantic level is analogous to sharing the schema or format specification at the syntactic level. This argument is plausible, except for the fact that ontologies are inherently extensible. Ontologies can be inherited from multiple, heterogeneous sources and are not prescribed in the way that syntactic format specifications are.

Another criticism of semantic approaches is that they generally rely on syntactic ones. Ontologies and description logics require a common data format and common protocol to structure their interchange—usually RDF serialized into XML. Fortunately, the syntactic requirements of languages for semantics are very low level and not likely to change.

So why hasn't syntactic or semantic interoperability solved the problems of CAD and GIS integration? There appears to still remain enough difference in the analysis approaches used to make complete interoperability possible. A useful exercise is to step away for a moment from structural approaches and to focus on the problems domains where such interoperability would be most useful. The next section presents several use cases that illustrate the most useful scenarios for flexible interoperability, based primarily on semantics and not on predefined interfaces.

7.3 Use Cases for Semantic CAD and GIS Interoperability

The functional divisions between planners, designers, and constructors have contributed to the separation of tools. But now as projects are conceived and delivered differently (e.g., design-build project delivery), with a whole lifecycle view from needs analysis to asset management, a common information platform is needed. It is worthwhile to present some use cases that illustrate the need for semantic interoperability and lead up to the later discussion of BIM as an enabling technology.

7.3.1 Facility and Infrastructure Planning

Although much work is still needed, some has been done to define the use cases for integration of facility and infrastructure planning. Peachavanish et al.[9] cite two compelling examples. In the first example, they discuss the need for geospatial data (i.e., location and soil strength) to support the design of a multi-story commercial building. For the design of a deep foundation for the building on a steep slope, data are needed about the adjacent structures' foundations. They correctly assert that such data is not easily obtained from a CAD or GIS system. Even a well maintained computer-aided facilities management (CAFM) or automated mapping/facilities management (AM/FM) system for the adjacent buildings would not likely hold the needed engineering data. Such data are typically obtained from the "as-built" plans and specifications or from geotechnical inspection reports. Had the engineering properties been captured in a machine readable semantic fashion, as in a geotechnical engineering ontology, they might be discoverable by a software agent.

The second use case discussed by Peachavanish et al.[9] involves the lack of precisely defined utility locations. Utilities not identified by site location efforts or through site plans of known locations can cause considerable delay and disruption in a project. A geospatial ontology for infrastructure utilities might capture, instead of the exact locations, the likelihood of certain infrastructures given the past uses of a site. For example, if the site is near an established easement or right-of-way, an *infrastructure agent* might be able to correctly infer the presence of buried utilities that can lead to more detailed sub-surface investigations, provided an ontology exists. Or, if a site formerly housed a commercial or industrial facility, an ontology might help site engineers determine what and, more beneficially, where to look to locate abandoned or undiscovered utilities.

7.3.2 Security Engineering

In March 2007, the OGC and Geospatial Information Technology Association (GITA) held the Emerging Technology Summit on the topic of Geospatial Web Services for Infrastructure and Buildings. A consortium of vendors and academics was invited to prepare their visions for CAD/GIS/3D/BIM convergence in a variety of application areas. One area that had some interesting opportunities for integration was security engineering. Security engineering for buildings combines procedures for fire safety, chem/bio/terror attack prevention, and surveillance requirements. As alluded to in the introduction to this chapter, this application is one clear example that combines the in-building and out-of-building perspectives.

As a contribution to the Summit, Young and Sankaran[18] presented a security engineering scenario with specific geospatial queries that span these perspectives. They include:

- How can we move a specific number of employees from a place of work considering geospatial constraints?
- How can we find entrance, exit locations, and fire extinguisher placements?
- How can we find the best place to locate security guards around facilities?
- How can we determine what a guard can see from a given location?

These questions lead to many other queries that can be developed that span the in building and site domains. What were not adequately addressed at the Summit were strategies for answering these queries. Interoperability seems necessary, but without agreement on CAD and GIS format data exchange, it is not clear how such interoperability will take place.

Semantic approaches, which will be described in the case study in Section 7.5, can take advantage of BIM as a CAD/GIS hybrid.

7.3.3 OGC Web Services Testbed Phase 4

OGC develops standards and specifications for sharing geospatial information by engaging information users, software developers, and people from academia in interoperability experiments and prototyping testbeds for Open Geospatial Web Services. These testbeds focus on establishing requirements for OGC services, specifications, and demonstration of interoperable applications that integrate information exchange between the geospatial and AEC communities.

The Open Geospatial Web Services Testbed Phase 4 is an interoperability experiment carried out for CAD-GIS-BIM integration that aims at developing requirements for bridging the information interchange and workflows between the AEC world and those of the geospatial community. The U.S. Geospatial Services Administration (GSA) and the U.S. National Geospatial Intelligence Agency (NGIA) are the two main contributors to this testbed. Their objective was to test Space Assessment capabilities of the National Building Information Modeling Standards with 3D visualization in a geospatial Web services architecture.[19]

There are many scenarios in which these Web services could prove beneficial. Considering the interest of participants in an urban environment, a disaster management scenario was chosen for the testbed. Disaster management requires specialists from diverse backgrounds to collaborate with accurate information in a short span of time. These specialists may be planners using GIS or CAD, and architects using BIM who are required to share their respective data with one another, thus creating a seamless integration of data for effective disaster management. Keeping in view the incompatibilities that exist between geographic and CAD data, such as geographic data usually being geo-referenced, while CAD/BIM data are not, as well as the native formats in which these exist, necessitates the requirement for Open Web Services. Using Open Web Services, client applications that can handle data in individual proprietary formats after seamless integration could be developed for delivering up-to-date information.

For the testbed, a fictitious scenario involving the explosion of a "dirty" (low-level radioactive) bomb in New York City was imagined. The objective of the testbed was to setup an emergency medical facility close to an airport installation. In order to identify the suitable site and to setup the temporary facility, it is first required to locate buildings close to an airport over a vast geographic area. Second, the egresses have to be planned so that they are sufficient enough for emergency activities that may be carried out at an emergency facility, and third, detailed drawings must be designed for setting up the field hospital.

The first step in the testbed scenario requires a planner to obtain information on building spaces of a large number of buildings spanning a vast area close to the airport. For this purpose, the planner makes use of a 3D client, such as the LandXplorer CityGML viewer. With the help of an OGC Catalog service, the planner is able to identify services that provide the required information, and thus retrieves building space attributes for several buildings. This is done via a Web Feature Service (WFS) for BIM and other city features such as road networks and terrain data from a WFS serving CityGML models. Once, all the necessary information is obtained and overlaid on top of one another, data is analyzed and reports are generated that enable the planner to make a decision. As soon as a building is identified to be suitable for the temporary hospital (which in this case is a hangar in the Newark airport) the planner registers the building information on an online public catalog, thus making it readily available to the next person responsible for managing the space in the building.

The airport hangar that is identified as the most suitable location for setting up a field hospital needs to be accommodated for multiple spaces such as a surgery room, patients's rooms and so on. For this task, the planner uses a CAD/BIM Editor and the WFS-BIM server from Onuma, Inc. (http://www.onuma.com). The information made available by the planner in the first stage is now requested by the space planner from the OGC catalog, which returns the Web service address of the hangar. The space planner can now request hangar information from the WFS, which is delivered in the form of IFCs. The planner now uses the CAD/BIM editor to create new spaces required for typical emergency activities and then returns this new space scheme to the server in IFC.

The availability and location of this data are stored in the OGC catalog service for use by the next collaborator. Once the planning of hospital spaces is done, the next step is to design the detailed drawings for construction of the facility. The designer uses MicroStation from Bentley Systems as the CAD/BIM editor client. He first obtains contextual information that would help him understand where the construction would take place. From the OGC catalog, he requests an aerial image of the area of interest using Web Map Service (WMS) and the surrounding city features in the form of CityGML from the found WFS. Lastly, he obtains the space schema of the field hospital with detailed rooms, developed by the space planner from the WFS-BIM server in the form of IFCs. With the data collected from the three resources, the designer uses architectural tools for 3D visualization of the hangar and generates detailed drawings, which are then presented for construction.

7.4 Building Information Modeling (BIM)

Building Information Modeling (BIM) is a technology platform that has emerged within the past five years that embodies the tools, methods, and standards for managing the facility lifecycle. There is irony in that the technology is much older than that, with the first planning for "intelligent CAD" dating back to 1975.[11] And yet, only recently has the vision that was conceived more than a quarter century ago finally been realized through steady advancements in 3D/4D (3D plus time) visualization, dynamic control, and parametric object modeling. But, perhaps most importantly, only now has the desktop hardware of today—with multi-core processors operating on the order of 4 GHz, several gigabytes of RAM, and advanced graphics cards—made the vision of BIM a reality.

It must be stated that BIM is really not new, but merely the result of a steady evolution of CAD, with its heritage deep in CAD for architectural design. But BIM is not just the new CAD. Rather it has adopted many "best of breed" technologies during its development, including many of the intelligent behaviors of GIS such as associated data and rules, non-redundant geometry (i.e., topology), and automatic modification of associated geometries, among others.[11] But, BIM has not become a close relative of GIS, either. BIM is still focused very much on buildings. In fact, commercial BIM applications are presently geared only toward targeted building design disciplines. Autodesk Revit®, for example, is currently only available in three versions: for architecture (Revit Architecture), structural (Revit Structure), and mechanical, electrical, and plumbing (Revit MEP). What distinguishes these products from Autodesk and similar products from other vendors (Vico, ArchiCAD, Bentley, etc.) are the objects and methods that come packaged with the application. At present, the objects for larger scale design disciplines such as civil infrastructure or site design are not available. As a result, the ability for BIM to interoperate with GIS, at least at present, is somewhat limited. This section will briefly discuss the current state of the practice with BIM, present a design sequence using one of the tools, discuss the standardization efforts currently underway, and discuss the future vision of interoperability with BIM using semantic methods.

7.4.1 BIM: State of the Practice

The AEC industry is at a watershed point in its long history and BIM is at the center. Trends that have been decades in the making are converging at the same time and leading to unprecedented change. Some examples include:

- **Focus on alternative project delivery methods.** Traditional design-bid-build is being replaced with design-build project delivery. Projects can be delivered faster and with better coordination when singular responsibility is placed with one design-build team. Project teams are incentivized to deliver innovation and creativity along with accelerated project schedules. BIM is the tool of choice for coordination of phased design and construction.
- **Focus on integrated design and management.** With BIM, it is possible for interdisciplinary design teams such as structural, fire safety, and heating, ventilation, and air conditioning (HVAC) engineers to coordinate their designs with the same model at the same time. It is also possible for integrated *planning* with facilities managers that manage geographically distributed assets such as campus environments, retail chains, and transportation/logistics firms. *This is the key trend that closely relates the need for geospatial integration with facilities modeling in CAD and now BIM.*
- **Focus on sustainability.** The green building movement has been around for decades, but with formal procedures now in place through the Leadership in Environmental & Energy Design (LEED) certification process, building owners have objective incentives for incorporating energy and waste saving measures into their projects. BIM technology allows easy auditing of design alternatives for sustainability.
- **Focus on asset management.** CAD was invented to automate the drafting process. CAD never really affected the paradigm of design, only supported it. Derivative technologies such as CAFM and AM/FM systems emerged to

provide tools for supporting the operations and maintenance of a building after it was built. With BIM, the potential for one information model to support the entire facility lifecycle is realized. The same model developed in design can be used to consider O&M and, in some instances, influence the design in ways that were never possible with CAD alone.

7.4.2 BIM: Tools and Methods

A lengthy discussion of the various BIM tools commercially available today is beyond the scope of this chapter. Besides the already mentioned Revit products, Table 7.3 presents a brief summary of the major commercial vendors, their main product(s), and application areas. Readers are encouraged to review Eastman et al.,[11] who provide an extensive treatment of the currently available tools, their capabilities, and limitations.

Table 7.3 Summary of Sample Commercial BIM Software and Primary Features/Capabilities

Vendor	Products	Features	Application Areas
Autodesk (www.autodesk.com)	Revit 2009 Architecture, MEP, Structural, NavisWorks	Integrated clash detection, design review	Architectural design, interdisciplinary coordination
Bentley (www.bentley.com)	Bentley Architecture, MEP, MicroStation, LEAP, RAM, STAAD, GEOPAK	Specialized engineering applications based on MicroStation and BIM platform	Building and functional system design, site design, transportation
GraphiSoft (www.graphisoft.com)	ArchiCAD 12	Object-database technology	Integrated, component-based building design
Tekla (www.tekla.com)	Tekla Structures	Focus on structural design and detailing	Widely used for structural fabrication, extensions for supply chain management
Vico Software (www.vicosoftware.com)	Constructor, Estimator, Control	5D virtual construction	Integrated Project Teams (IPT)

What is valuable to discuss here is the general sequence for how BIM models are developed and used. BIM begins with parametric objects with behavior that governs their interaction with other objects. 3D design tools that do not enable behavior are not BIM tools (e.g., Google Sketchup). The parametric objects used for building design can come from a variety of sources including material owners, vendors, software vendors, or standardization bodies. Base-level objects are typically provided by the vendor. For example, Autodesk Revit Architecture provides generic base-level "families" for building components such as walls, doors, and windows. These can be extended to add new properties or behaviors in keeping with a particular designer's established practices. Commercial-level objects can be supplied by a material vendor such as a door and window manufacturer. Client-level objects can be supplied by a large owner/client such as the federal government with an interest in standardization for its building aesthetics and performance criteria.

What follows the availability of a repository of objects is a direct, 3D designing process. Instances of objects are placed in a 3D "canvas" and refined according to the control parameters of that particular object. New objects are added and integrated with one another until a built-up assembly is created. Gross geometries (shapes) are created and then trimmed to create the desired final object. Assemblies are combined, and eventually, a basic structure is developed. The initial model can be subsequently shared with the integrated design team to allow concurrent and coordinated design to occur.

Beyond the basic modeling process (which can be iteratively refined over time by a team of designers from different disciplines), BIM allows conflict detection and constructability review. Building components that clash spatially (e.g., an HVAC duct intersecting with a steel beam), or by their performance characteristics, are identified and can be resolved systematically. In pre-construction activities, contractors and designers can develop construction sequences (i.e., 4D models) and plans for temporary shoring, crane placement and accessibility, and material staging. This process is generally referred to as virtual construction and design.

Generally speaking, BIM is a design communication medium. It allows a digital model to be developed collaboratively with all stakeholders' input. It encodes explicit design intent in a manner that all participants can access, refine, and adhere to.

7.4.3 BIM Standards

In December 2007, the U.S. National Institute of Building Sciences (NIBS) released Version 1 of the National Building Information Modeling Standard (NBIMS).[19] The goal of NBIMS is to promote standardization for the use of BIM for federal facilities and to promote a forum directed toward updating and obtaining consensus changes to published standards. NBIMS is being released in multiple parts, with only Part 1 available at the time of this writing. NBIMS v1 Part 1 is divided into several topics, primarily related to information exchange.

Besides NBIMS, there are no other standards that are widely adopted. To address the requirements of BIM for the design-build industry, the Design-Build Institute of America (DBIA) is promoting guidance for project owners on the specification of BIM requirements for design-projects. Other professional and advocacy organizations are pursuing similar efforts. Technologically speaking, the software is not yet mature enough to define, much less adopt uniform standards for especially syntactic ones. The opportunity for research in semantic means for standardization is open.

7.4.4 BIM Interoperability

The history for interoperability with BIM is for the most part shared with CAD; except of course, that low level data interchange formats such as DXF are not typically used with BIM. The elemental BIM objects are building elements as defined by product models such as the IFCs. Surprisingly little has been done so far to promote interoperability between BIM packages. The objects tend to be defined at such a high level of abstraction that interoperability is difficult. Not every commercial package provides all the features and capabilities needed by all users, so users choose the products that best meet their needs. Interoperability with CAD and GIS tends to be even more remote. Herein lies the opportunity for semantic interoperability with BIM, which is discussed in the form of a case study in the next section.

7.5 Case Study: Semantic BIM-GIS Integration for Campus Emergency Management

The tragic shootings at Virginia Polytechnic Institute and State University (Virginia Tech) in April 2007 have forced colleges and universities nationwide to put increased scrutiny on their emergency management procedures. The biggest criticism of the crisis managers after the Virginia Tech shootings was the delayed notification sent to students and staff apprising them of the magnitude of the situation, and the lack of sufficient instructions for evacuation or for finding refuge.

As was also the case with many higher educational institutions nationwide, George Mason University instituted a comprehensive alert system called "MasonAlert" in the fall of 2007 to distribute alert information through email, telephone, and SMS (Short Message Service) text messages in the event of an emergency.[20] The system permits police, first responders, and administrators to rapidly distribute emergency messages to students across multiple devices with alert messages and instructions. The messages tend to be of a broadcast nature with the goal being timeliness of notification, not precision of instructions.

But precision of instruction is an important consideration when different instructions need to be sent to different campus populations (i.e., students vs. staff) or locations of the campus in real time. The need to manage evacuation routes, in-building and out-of building hazards, and the demands of peak classroom attendance, all contribute to the need for advanced precision for these alert systems with a new emphasis on the interoperability of the information systems that support them.

The information systems to support crisis response are starting to become available. Some are GIS-based for representing campus assets and some are CAD-based (or CAFM-based) for use in buildings. What is apparent is that an integrated approach is not available to easily manage the *transition* from inside the building to outside. GIS and geospatial analysis does not have built-in methods to analyze spatial relationships such as "around the corner from." If structural damage occurs, conditions may not exist to locate a person or asset on a particular floor.

Ontologies and product data specifications are well defined for buildings, and object models (and some ontologies) are used for GIS. Where BIM can come into play is that the object families that make up the "building blocks" of BIM have specified structure and object properties and behavior. These objects can, in effect, *subscribe* to ontologies that will allow more subtle conditions and constraints to be modeled and the object families themselves to be extended. This is a very important point and goes to the heart of the matter with regard to semantic interoperability. The objects and their respective interfaces may remain distinct while allowing external relationships, rules, and behaviors to be applied by subscription to an appropriate semantic service.

7.5.1 Case Definition—Semantically Integrated Evacuation Planning

We present a fictional, though realistic, scenario of a campus evacuation scenario. The police (emergency response) office of a university receives an emergency call from a student claiming that a deranged individual is discharging a firearm inside a multistory classroom building. Besides reporting the incident, the terrified student asks immediately for instructions on how to evacuate the building safely, and how to reach a designated refuge building on another part of the campus. Sounding an all-building alarm, or broadcasting an electronic alert message to all students expected to be in that building at that time, may put lives in danger. How can disparate CAD (BIM) and GIS information systems aid an emergency response operator receiving this call to provide timely and precise information? This scenario brings into focus the need for dynamic evacuation modeling that can model both inside and outside building conditions. The scenario is depicted in Figure 7.4. The armed attacker is suspected to be moving between floors and would interfere with the closest known exit.

It is safe to assume that floor plans for all campus buildings would be available to an emergency response call operator. The likely medium for these plans is in CAD or, more likely perhaps, image files. For this building, we will further assume that a complete BIM is available with constituent components assembled from IFC objects. Assuming that the distressed caller can identify their location (floor, nearest room number), the call

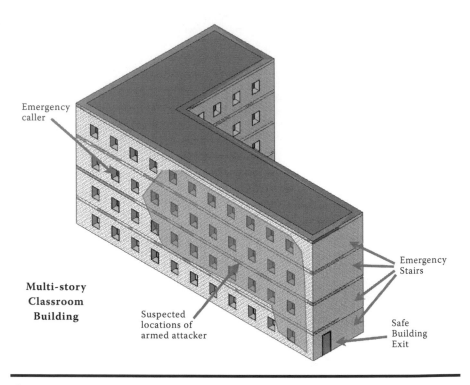

Figure 7.4   **Schematic of a multi-story campus building under siege.**

operator's immediate need is the planning of a route to an exit, and then a safe path once outside the building. But which exit is the safest one? If similar calls arrive from other threatened students in *different* parts of the building, then paths to certain exits may make the nearest one, for example, more dangerous. The scenario may quickly become too complicated for an operator to analyze. The support of an information system to provide the caller with the best evacuation route as quickly as possible is desired.

We propose a semantic route service that can dynamically assemble evacuation instructions, in the form of an SMS text message, to be relayed to the distressed caller's mobile phone. The service is unique in that it: (1) is targeted to an individual, not a mass-broadcast, (2) takes into account concurrent threats with dynamic avoidances, (3) analyzes the internal structure of the BIM model of the building, (4) plots an internal evacuation route to an appropriate exit, and (5) extends the evacuation route to the designated refuge point across campus. The service accomplishes this using semantic integration of the available BIM data from inside the building and GIS data from the campus road and pedestrian paths.

A multi-floor classroom building, according to most fire and life safety codes, requires stairwells suitable for evacuation. Stairwells are defined within a hierarchy of IFCZone, IFCSpace, and finally IFCStair. The assigned attributes include a description, geometrical representation, and set membership (cardinality) within or outside of other collections of objects.

A semantic service is capable of extending the IFC definitions of rooms, corridors, stairwells, and entry doors in a way that facilitates in-building navigation. In addition to gathering the geometrical representations of the building components and set membership, the semantic service can dynamically assemble constraints such as blocked doors or threatened areas. Using OWL Lite, we can encode a simple instance of an ontology that extends the IFC definition of a stairwell in this manner. The instance is combined with an instance of a navigation ontology in order to synthesize the mechanics of an internal route plan as shown in Figure 7.5. Because the stairwell is modeled as a fully 3D BIM (IFC) object, the extended geometrical and relationship properties can be queried by the semantic service. This would not have been possible given only a vector CAD representation based on elemental geometry.

Once outside the building, the semantic service draws from GIS databases to determine an overland route via campus roads and pedestrian walkways to the

Semantic proposition: Stairwell all of whose exits are either more stairs (unblocked) or have an outdoor exit:

OWL Markup

```
<owl:Class>
  <owl:IntersectionOf rdf:parseType="collection">
  <owl:Class rdf:about="#Stairwell"/>
  <owl:Restriction>
    <owl:onProperty rdf:resource="#hasChild"/>
    <owl:allValuesFrom>
    <owl:unionOf rdf:parseType="collection"/>
      <owl:Class rdf:about="#Stair"/>
      <owl:Restriction>
        <owl:onProperty rdf:resource="#hasChild"/>
        <owl:someValuesfrom
  rdf:resource="#Stairwell"/>
        <owl:Restriction>
        <owl:unionOf>
        <owl:allValuesFrom>
      <owl:Restriction>
    <owl:IntersectionOf
  <owl:Class>
```

IFC definition (Sample)

Attribute	Relations	Definition
Vertically Connects	LIST [0:?] OF IfcSlab	List of Floors to which this stair assembly connects. Through these relationships, one can determine which building stories are served by this stair.

Figure 7.5 Relationship between stairwell IFC definition and OWL semantic description of logical test for blocked exits.

refuge building. The service is capable of discovering the network topology of available paths as well as impedances. The semantic service subsequently formats an appropriate, targeted evacuation plan in the form of an SMS text message to the mobile phone of the emergency caller.

Admittedly, this case illustration is conceptual and based on numerous assumptions. Life safety is a critical need, but established protocols in use by emergency first responders will take precedence over any automated evacuation response system. Still, this approach does offer a window into the types of integration that may be possible in the future between CAD (BIM) and GIS to support emergency decision making. The section that follows discusses some extensions that might increase the capability of the suggested semantic evacuation service.

7.5.2 Considerations for Advanced CAD (BIM)/GIS Semantic Integration

Here are several considerations that would need to be addressed in order to implement an advanced CAD (BIM)/GIS semantic integration service for evacuation management:

■ **Building occupancy loads and congregation points.** Campus buildings undergo different levels of occupancy throughout the day. Depending on the level of occupancy, evacuation instructions should take into account the capacity for stairwells and other egress points to accommodate variable crowds.

■ **Building sight lines.** In the case of an armed gunman, the sight lines in between buildings may be important to not route evacuees to locations in range of fire. Such considerations are needed from both inside and outside the building.

■ **Strategic broadcast and targeted SMS messages.** How can semantic services dynamically combine wide-release broadcast messages with targeted, individual instruction-level messages? For example, can a security camera picture of the perpetrator be sent to all students in conjunction with targeted messages? It would also be desirable to send priority messages to faculty and staff based on their assigned office spaces.

■ **Predictive approaches.** Advanced semantic services may be capable of inferring the location or trajectory of the perpetrator. Sometimes office lock-down is preferable to evacuation.

At this point validation of the approach is not practical, as no specific implementation of BIM-based ontologies or examples of semantic integration with BIM exist. Nonetheless, this case study has provided a useful illustration of the wide scope of potential integration that will likely become possible for resolving complex crisis management situations that rely on integration of multiple data sources, at multiple scales, and in a real-time situation.

7.6 Chapter Summary

This chapter has presented a broad view of CAD/GIS integration from the perspective of semantics, and profiled the use of BIM as an enabling technology. The chapter has compared and contrasted syntactic vs. semantic interoperability and the advantages and disadvantages of each; profiled use cases for semantic interoperability related to evacuation modeling in campus environments; discussed Building Information Modeling as a technology platform capable of bridging CAD and GIS; and described a case study on semantic processing of CAD and GIS data in a BIM application for evacuation.

BIM is a powerful technology that can automate some tedious and repetitive parts of the design process and allow buildings to be designed in a truly integrated manner. BIM is leading to a paradigm shift in the way building projects are conceived, planned, design, built, and maintained.

7.7 Conclusions

Unlike syntactic approaches, a first tenant of semantic integration is that prior agreement on interfaces for disparate information systems are not necessary. This holds true, and is increasingly necessary, for integration between CAD and GIS systems. This chapter has attempted to discuss the current state of commercial CAD and GIS software with respect to semantic integration, and in particular the role that BIM can play in enabling this integration. Through the provided case study, a simple case for emergency evacuation has shown how semantic services can capture not just the geometry of CAD (BIM) and GIS objects, but also higher level concepts such as blocked doors expressed in the form of semantic Web languages

One point about ontological representations that should be made is their inherent openness. Ontologies on open networks such as the WWW have no authoritative sources. In other words, facts derived from instances of an ontology can only be proven to the extent of the statements given. They may be logical, but not physically correct. To properly validate information systems for emergency response situations, this degree of openness may pose a concern. Even with semantic approaches, the correct encoding of domain knowledge is still required. As is the case with most technological development, validation and security usually lags innovation and the semantic integration of CAD and GIS is no exception.

References

1. Berners-Lee, T., O. Lassila, and J. Hendler. "The Semantic Web." *Scientific American*, May 2001.
2. W3C. "Resource Description Framework (RDF)." *World Wide Web Consortium (W3C)*. 1999. http://www.w3.org/RDF/ (accessed March 2008).
3. IAI. "IFC/ifcXML Specifications." *International Alliance for Interoperability (IAI)*. February 2006. http://www.iai-international.org/ (accessed March 2008).

4. ISO. "ISO 10303: Industrial automation systems and integration—Product data representation and exchange." 2003.

5. ISO. "ISO 10303-11: Industrial automation systems and integration—Product data representation and exchange—Part 11: Description methods: The EXPRESS language reference manual." 2004.

6. ESRI. "ESRI Data Models." *Environmental Systems Research Institute.* 2004. http://support.esri.com/index.cfm?fa=downloads.dataModels.matrix (accessed March 2008).

7. OGC. "Geography Markup Language." *OpenGIS Consortium.* 2007. http://www.opengeospatial.org/standards/gml (accessed March 2008).

8. W3C. "W3C Geospatial Ontologies." *World Wide Web Consortium.* October 2007. http://www.w3.org/2005/Incubator/geo/XGR-geo-ont-20071023/ (accessed March 2008).

9. Peachavanish, R., H. Karimi, B. Akinci, and F. Boukamp. "An ontological engineering approach for integrating CAD and GIS in support of infrastructure management." *Advanced Engineering Informatics* (Elsevier) 20 (2006): 71–88.

10. Scacco, M.J. "Changing the landscape: 3-D CAD modeling is emerging as the standard civil engineering design solution." *CE News.* February 2008.http://www.cenews.com/article.asp?id=2773 (accessed February 2008).

11. Eastman, C., P. Teicholz, R. Sacks, and K. Liston. *BIM Handbook: A Guide to Building Information Modeling for Owners, Managers, Designers, Engineers and Contractors.* Wiley, 2008.

12. Autodesk. "AutoCAD 2009 DXF Reference." *Autodesk.* January 2008. http://images.autodesk.com/adsk/files/acad_dxf.pdf (accessed March 2008).

13. ESRI. "ESRI Shapefile Technical Description." *Environmental Systems Research Institute (ESRI).* July 1998. http://www.esri.com/library/whitepapers/pdfs/shapefile.pdf (accessed March 2008).

14. ESRI. "ArcGIS Data Interoperability." *Environmental Systems Research Institute.* March 2008. http://www.esri.com/software/arcgis/extensions/datainteroperability/index.html.

15. National Institute of Building Sciences (NIBS). *US National CAD Standard v4.* January 2008. http://www.nationalcadstandard.org/index.php (accessed March 2008).

16. OGC. "Web Feature Service (WFS)." *OpenGIS Web Feature Service (WFS) Implementation Specification.* 2004. http://www.opengeospatial.org/standards/wfs (accessed March 2008).

17. Casey, M., and A. Austin. "Semantic Web methodologies for spatial decision support." *DSIAge2002: International Conference for Decision Support Systems in the Internet Age.* Cork, Ireland, 2002.

18. Young, J., and S. Sankaran. "CAD/GIS/3D/BIM Convergence: ESRI Support." *OGC/GITA Emerging Technology Summit.* Washington, DC, 2007.

19. National Institute of Building Sciences (NIBS). *United States National Building Information Modeling Standard, Version 1—Part 1, Overview, Principles, and Methodologies.* Facilities Information Council, National Institute of Building Sciences, 2007.

20. George Mason University,. *Mason Alert.* 2007. https://alert.gmu.edu/ (accessed April 2008).

21. Yuan, L., and H. Zizhang. "3D indoor navigation: A framework of combining BIM with 3D GIS," 44th ISOCARP Congress, 2008.

22. Shayeganfar, F., A. Anjomshoaa, and A. Tjoa. "Smart indoor navigation solution based on Building Information Model and Google Android" *LNCS: Computers Helping People with Special Needs*, Volume 5105/2008, Springer-Verlag.

Chapter 8

Ontologies for Linking CAD/GIS

Tamer El-Diraby and Hesham Osman

Contents

8.1 Introduction

Within the plethora of available tools to achieve interoperability between computer-aided design (CAD) and geospatial information systems (GIS), ontologies offer some attractive benefits. Standardization, product models, and data exchange standards have all contributed to alleviate the CAD/GIS interoperability dilemma, albeit all at the level of *data* and/or *information*. Within the domain of interoperability, ontologies are attractive as they contribute to a much deeper level of interoperability, that of *knowledge*.

The term *ontology* originates from philosophy as it refers to the branch of metaphysics concerned with identifying, in the most general terms, the kinds of things that actually exist, and how to describe them. This mind-set of *conceptual modeling* appealed to computer scientists in the mid-1980s and early 1990s, when extensive research was conducted into the use of ontologies as enablers of software interoperability and repositories for knowledge representation. Interest in ontologies resurged in the late 1990s with the advent of the Internet and the prospects of achieving a "Semantic" Web where knowledge, rather than data, is shared among both humans and intelligent software agents.

Linking CAD and GIS systems poses challenges that can be addressed through the use of ontologies. Unlike product models and data exchange standards that specify *how* various concepts should be described and formatted, ontologies can specify *what* each concepts means, its *relationship* to other concepts and *rules/axioms* that govern the way various concepts interact.

8.2 Ontologies: A Subtle Introduction

The term ontology originates from philosophy. In that context, it is used as the name of a subfield of philosophy, namely the study of the nature of existence (the literal translation of the Greek word Οντολογια), the branch of metaphysics concerned with identifying, in the most general terms, the kinds of things that actually exist, and how to describe them [1]. However, in more recent years, ontology has become one of the many words hijacked by computer science and given a specific technical meaning that is rather different from the original one. Typically an ontology consists of a finite list of terms and the relationships between these terms. The terms denote important concepts (classes of objects) of the domain. For example, in the domain of highway construction, *Designers*, *Contractors*, *Equipment*, *Payment*, and *Specifications* are some important concepts. The relationships typically include hierarchies of classes (a *Dozer* is a subclass of *Equipment*). Apart from subclass relationships, ontologies include information such as the following:

- Properties (*Designers* produce *Specifications*)
- Value restrictions (Only *Qualified Personnel* are allowed to supervise particular construction activities)

- Disjoint statements (*Contractors* and *Designers* are disjoint)
- Axioms (If *Construction* has been completed as per *Specifications*, then *Contractor* is entitled to *Payment*)

Some of the differences between ontologies and alternative data/information representation tools that are widely used include the following [2]:

- The language for defining the ontology is syntactically and semantically richer than common approaches for databases.
- The information that is described by an ontology consists of semi-structured natural language texts and not tabular information.
- An ontology must be a shared and consensual terminology because it is used for information sharing and exchange.
- An ontology provides a domain theory and not the structure of a data container.

It has been argued that the semantic structuring achieved by ontologies differs from the superficial composition and formatting of information (as data) afforded by relational and XML databases [3]. Ontologies are able to provide an objective specification of domain information by representing a consensual agreement on the concepts and relations characterizing the way knowledge in that domain is expressed. This specification can be the first step in building semantically aware information systems to support diverse enterprise, government, and personal activities. In brief, ontologies are considered to be *knowledge-models* rather than *data/information models*. In this sense, ontologies can be considered as the main facilitators of (1) re-use of knowledge, (2) ordering and structuring of knowledge, and (3) analyzing of knowledge.

Ontologies are usually created in a layered architecture whereby the more specific inherits (or maps) concepts from the more general ontologies [4]. This allows for reuse of generic concepts created at the top layers and assists designers of lower level ontologies to focus on their specific domain of knowledge. In this regard, a distinction is made between four levels of ontologies:

1. Fundamental/Top-level Ontologies: These include very rudimentary concepts that are common among all domains of knowledge. Examples include the Cyc Ontology and the Standard Upper Ontology [4].
2. Domain-level Ontologies: These build on the basic concepts defined by fundamental ontologies. Domain ontologies define those general concepts and relationships that are inherently specific to a particular domain of knowledge. Examples of domain ontologies could include ontologies in the fields of medicine like USMLS [5], tourism like Onto-Tour [6], or transportation. Development of these ontologies should be a cross-industry effort.
3. Application-level Ontologies: These build on the concepts defined at the domain-level but are intended for use in a particular use-case scenario. Examples of application ontologies related to transportation could include

ontologies in the fields of highway geometric design, context-sensitive design of urban streets, or regional transportation planning. Development of these ontologies is usually a cross-industry or inter-organizational effort.

4. User-level Ontologies: These can be considered to be the most task-specific kind of ontologies that are created to cater for a specific application within a particular organization. Examples of potential application ontologies include ontologies for utility coordination within a particular city, supply chain management for a particular highway construction project, or bus scheduling within a particular region. Development of these ontologies is usually an intra-organizational effort.

8.2.1 Ontologies and Cascading Levels of Interoperability

A recent study by the National Institute of Standards and Technology estimates the cost of inadequate interoperability among computer-aided design, engineering, and software systems in the U.S. capital facilities industry to be $15.8 billion per year [7]. The term *interoperability* is used to refer to the ability of systems or—in the boarder sense—organizations, to communicate in a collaborative environment. That being said, interoperability can be assumed to exist at three cascading levels of complexity. At the lowest level, data interoperability is concerned with achieving the ability to exchange data across different systems. This level is mainly concerned with low-level file format issues and data representation consistencies. At a higher level, information interoperability is mainly concerned with the ability to interpret and understand the meaning of data that is being exchanged. In this regard, metadata standardization initiatives like the Industry Foundation Classes and Geography Markup Language [8] fall along the lines of information interoperability initiatives. Finally, the highest level of interoperability is that of knowledge, whereby systems and organizations not only exchange and interpret information, but also are able to deduce new information that is not explicitly defined. Ontologies can be considered to be one of the facilitators of knowledge interoperability.

The discussions pertaining to the use of ontologies for CAD/GIS integration in the following sections pertain to the elements of information/knowledge interoperability. The difference between ontologies and data models can be best demonstrated through an example. The following sections present an ontology for utility infrastructure products and their related concepts. Throughout the presentation of the ontology, issues pertaining to the discrepancies in design methodology, specification, implementation, and use of ontologies and data models will be highlighted.

8.2.2 An Ontology for Infrastructure Products

An *Infrastructure ProDuct Ontolgoy* (IPD-Onto) has been developed at the University of Toronto as part of a much larger ontology development endeavor that will encompass far more concepts pertaining to infrastructure development. This initiative was conducted in close collaboration with the City of Toronto and various stakeholder utilities involved in the city's Utility Coordination Committee. Other ontology development efforts that are currently underway pertaining to processes and actors overlap with IPD-Onto. This section will focus mainly on IPD-Onto and will occasionally refer to the other ontologies where necessary.

Figure 8.1 depicts the layered architecture of the domain IPD-Onto and several potential application ontologies that can be built upon the concepts defined in the domain ontology. The application ontology discussed in this section is that for urban infrastructure routing (IPD-ROUTE). Software systems (or user-level ontologies) built on this ontology can be tailored to the needs of different organizations (utilities, local government, transportation agencies, or emergency agencies).

The layered architecture presented in Figure 8.1 reaps two main advantages of ontologies: knowledge reusability and interoperability [2]. For example, assume the potential user of the ontology (a major city) agrees to commit to a common domain ontology for infrastructure products, creates two application-specific ontologies that support infrastructure construction coordination and emergency management, and hence develops supporting software based on these task-specific ontologies. Several advantages would be met. First of all, the basic domain knowledge pertaining to infrastructure products would need to be defined only once and reused numerous times for the different applications. Secondly, the domain knowledge will be consistent among the different application domains, thus enabling heterogeneous systems within the city (or between the city and other organizations) to seamlessly interoperate. In the given example, one could envision a crisis situation where emergency management systems would require real-time information pertaining to the construction status of ongoing projects within the city. Current database-driven systems that do not share a common information representation create isolated "information islands" that hinder interoperability and create so-called "bottlenecks" in information flow.

IPD-Onto is composed of three layers. The first is the abstract root layer that contains a meta-model of high-level concepts modeled closely around the e-Cognos ontology [9]. The second and third layers are both domain ontologies that are created at different levels of detail. IPD-Onto Lite is the light-weight version of the ontology. IPD-Onto Full reuses the concepts defined in IPD-Onto Lite and builds on these concepts extensive product and attribute taxonomies specific to each utility infrastructure sector. IPD-Onto Lite contains approximately 25% of the concepts in IPD-Onto Full. The separation of the ontology into these two layers was intended to:

- Facilitate the integration process with other ontologies (process ontology and actor ontology) currently being developed in parallel
- Allow the upper-level models of products and attributes (defined in IPD-Onto Lite) to be used independently of the sector-specific taxonomies that were developed based on common Canadian terminologies (defined in IPD-Onto Full).

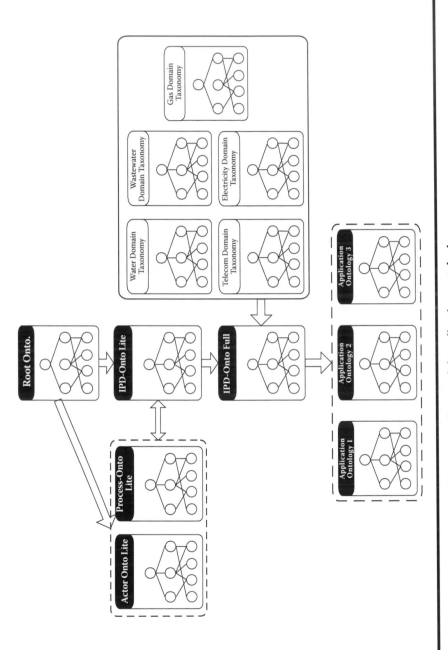

Figure 8.1 Layered architecture of domain and potential application ontologies.

In the aforementioned example, data exchanged between the city's emergency management and construction management systems would be virtually impossible if they did not subscribe to a common information model. The proposed layered ontological architecture would be able to bridge this interoperability gap.

8.2.2.1 Ontology Development Methodology

IPD-Onto used best-of-breed approaches for ontology development by drawing upon several approaches for conceptual modeling and evaluation. The following list the main four methodologies that were employed:

1. Requirement analysis (RA) was accepted as a vital initial step in system development [10]. RA served to act as the motivating scenario for the competency questions that the ontology needed to address. The RA was built on:
   - Analysis of product modeling requirements of stakeholders (abstract specification for domain ontology)
   - Analysis of the utility coordination requirements of stakeholders (concrete specification for application use-case)
2. Benchmarking existing models: Several information models for utility infrastructure products in the water, wastewater, telecommunication, gas, and electricity domains were reviewed and analyzed. The purpose of this benchmarking process was to:
   - Identify the strengths and weaknesses of each modeling initiative against the requirements set forth by the RA.
   - Reuse some of the concepts that were proposed by product models in their individual domains wherever possible.
3. Design Guidelines: An extensive review of design guidelines that govern the spatial constraints and requirements for buried infrastructure was performed. The design guidelines encompassed all five major utility sectors. Knowledge was compiled from various industry codes, standards, and manuals. Knowledge from these guidelines represents explicit constraints that govern infrastructure routing.
4. Case Studies: Eight major infrastructure projects in Southern Ontario were investigated from September 2004–August 2005. Seven of these projects involved municipal infrastructure (water/wastewater), while one project involved a highway interchange reconstruction. The case studies were specifically helpful in guiding the development of the application-level ontology and the routing decision models. The case studies were used to:
   - Identify some of the competency questions that should be addressed by the ontology.
   - Identify the main attributes of infrastructure products that impact the design process.

- Better understand the main issues facing designers during the process of design coordination.
- Identify and refine the main criteria that are usually considered during the routing of buried infrastructure at both the micro- and macro-levels.

8.2.2.2 Infrastructure Product

Physical products that are modeled in IPD-Onto include infrastructure products and street products. Infrastructure products include all physical products that belong to the main five sectors of utility infrastructure. Street products act as an umbrella for all non-utility products that can be found within the vicinity of utilities. These include landscape products (e.g., tress, shrubs, planter boxes), street furniture (e.g., benches, bicycle stands, telephone booths), and road features (e.g., curbs, medians, crosswalks, driveways).

IPD-Onto focuses on utility infrastructure products. An infrastructure product meta-model is proposed to represent higher-level concepts pertaining to infrastructure products. This upper meta-model serves as the core enabler for interoperability across various utility sectors. Infrastructure products can be viewed as being either generic or sector-specific products. A generic product is a product that is not unique to any specific utility sector (for example a generic pipe, manhole, or valve). A sector-specific product can be considered a manifestation of a generic product in a particular utility sector (for example a gas pipe, wastewater manhole, or a water valve).

Generic products can be grouped based on the role they play within the utility network. Grouping is made into one of three categories: main structures (e.g., lines, pipes, or cables), support structures (e.g., manholes, pedestals, or fittings), and devices (e.g., valves, meters, or sensors). Generic products can also be grouped based on their intrinsic function into one of seven functional groupings: conveyance, storage, measuring, control, protection, access, and locating.

At the sector-level, products can be found to exist at three levels of composition. System-level products represent the highest level of product aggregation and include high-order network entities like a complete distribution systems and entire water purification plants. Products at this level are composed of subsystem-level products (for example a water distribution system is composed of waterlines, pumping stations, and tanks). Subsystem-level products are in-turn composed of component-level products (for example a pumping station is composed of pumps).

In summary, the multi-modal nature of infrastructure products can be better explained through the following four orthogonal dimensions:

- Hierarchical role
- Sector
- Function
- Composition

Hierarchical role: This dimension is intended to classify a generic infrastructure product based on the role it plays within the infrastructure network. A generic product is assumed to be either a main structure, supporting structure, or device. Main structures are the basic linear carriers of the medium being conveyed by the infrastructure carrier. Supporting structures and devices provide auxiliary roles in the network to main structures. Devices are mostly mechanical or electronic objects that are designed to fulfill a specific purpose, whereas supporting structures are non-mechanical/electronic objects that are used to maintain the main structures.

Sector: This dimension identifies the sector-specific nature of infrastructure products by classifying infrastructure utilities into one of five main sectors: gas, electricity, telecommunication, wastewater, and water.

Function: This group classifies infrastructure products based on the functions each product performs. In this regard, seven functional categories have been identified:

Conveyance Products: Act as the direct carriers of the medium being supplied by infrastructure systems. Examples include gas lines, water lines, electricity cables, and fiber optic cables.

Control Products: Regulates the medium being supplied by the infrastructure system. Examples include gas valves, electric switches, and water control valves.

Protection Products: Provide protection to other infrastructure products or to entities in the vicinity of the product. Examples include cathodic protections devices, concrete duct banks, electricity ground points, relief valves, and manhole covers.

Access Products: Provide access to other infrastructure products. Examples include various types of junction boxes, manholes, and chambers.

Measuring Products: Perform some sort of attribute measurement to the infrastructure product itself (e.g., corrosion sensors) or to the medium being carried by the product (e.g., electricity meters, flow measuring devices, pressure gauges, and thermometers).

Storage Products: Accumulate the medium being supplied by the infrastructure product. Examples include water tanks, capacitors, batteries, and gas storage facilities.

Locating Products: Identify the location of buried infrastructure. Examples include markers and tracer wires.

Composition: This group is intended to capture the notion of aggregation and composition between products. The relationship between the three levels of products is a *composed-of* relationship.

Component products: These products represent the lowest level of infrastructure product aggregation. In essence, these products cannot be decomposed into any further products that lie within the taxonomy. Examples include individual pipe segments, fittings, meters, manhole covers, pumps, and electricity cables.

Sub-system products: This group encompasses all intermediate-level products that do not belong to the aforementioned groups. Examples include, water lines (that are composed of pipes, fittings, and valves), electricity lines (that are composed of cables and couplings), and pump chambers (that are composed of pumps, fittings, meters, and valves).

System products: This group is used to represent the top level of infrastructure product aggregation and include most high-order network entities. Examples include water distribution systems, storm water collection systems, electricity substations, and gas distribution systems.

An example of how some infrastructure products fit these proposed modalities is presented in Figure 8.2. The figure includes two main parts: the upper part shows the generic product model and the lower part shows the sector-based product model. The upper part illustrates how the generic products are modeled in two dimensions: "hierarchical role" and "function." The lower part includes two dimensions: the sector and composition. The relationship between the generic and sector-specific products is shown by the arrows: a Gas Valve is a Valve and a Wastewater Manhole is a Manhole. Similarly, subsumption relationships between compositional products are shown. For example, a Gas Service Line is composed of Gas Valves, Gas Pipes, and Gas Fittings.

8.2.2.3 Infrastructure Product Attributes

Attributes are considered to be one of the most important concepts that are needed to effectively describe infrastructure products. The model for infrastructure product attributes identifies two distinct dimensions of attributes: attribute types and modalities. Attribute types are clustered into eleven main groups (dimension, spatial, material, shape, cost, performance, soil, dependency, redundancy, state, and impact). Orthogonal to these eleven basic groups, additional modalities of attributes were generated (Figure 8.3). Six main modalities for IPD attributes were found to be of relevance.

- *Physical modality:* Physical attributes are usually tangible (e.g., diameter, width, material), whereas non-physical attributes include things like cost, performance, and dependency.
- *Change modality:* Fixed attributes are those attributes that do not change throughout the lifecycle of a product (e.g., shape, diameter, and material). On the other hand, changeable attributes assume various values throughout the product's lifecycle (e.g., performance, state of operation, and cost).
- *Phase Modality:* Used to identify the specific project lifecycle phases when a particular attribute is of relevance, for example, design attributes vs. construction attributes.

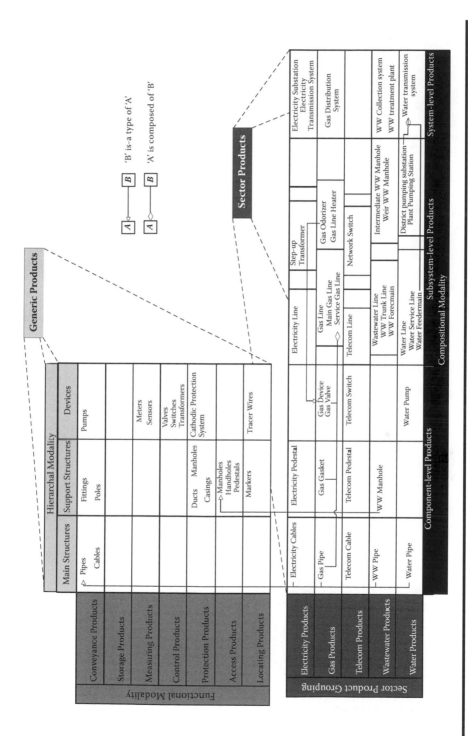

Figure 8.2 Examples of actual utility infrastructure products mapped in IPD-Onto.

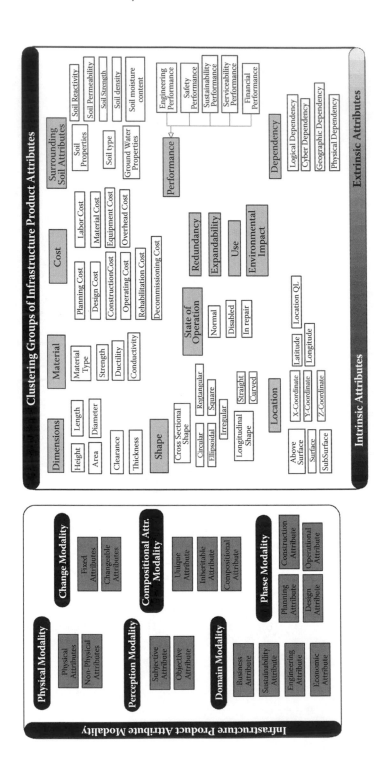

Figure 8.3   Clustering groups and modalities of some pertinent IPD attributes.

- *Perception Modality:* Subjective attributes are attributes that can have more than one appraising value that depends on the point of view of the actor, while objective attributes have values that are usually not contested.
- *Domain Modality:* IPD attributes usually tend to belong to one or more domains of interest. For example, environmental impact will tend to belong to the sustainability domain, whereas surrounding soil conditions are clearly of interest in the engineering domain.
- *Composition Modality:* This clustering concept relates to the compositional groups discussed in the previous section. This "meta-attribute" of attributes classifies how the values of attributes are inherited from component to subsystem products. In this regard, three compositional categories exist:
  - *Unique attributes*: These attributes are not inherited upwards in an aggregation relationship. Examples include performance and location. For example, the depth of a water pipe cannot be inherited upwards to indicate the depth of an entire water line.
  - *Inheritable attributes*: These attributes are inherited upwards in an aggregation relationship. Examples include surrounding soil attributes and ownership attributes. For example, if a water pipe is surrounded by sandy soil, then it can be inferred that the water line it belongs to is also surrounded by sandy soil.
- *Compositional attributes*: These attributes are combined together in an aggregation relationship. Examples include cost and length for pipes. For example, the cost of a transformer station is the sum of the cost of its constituent products and the length of a sewer line is the sum of the lengths of the pipes it is made up of.

8.3 Semantic Wrapping Using Ontologies

In order to demonstrate the interoperability capabilities of the IPD-Onto, a use case scenario is developed. The use case revolves around the process of utility infrastructure product routing in the urban environment. Simply put, the problem involves selecting the optimum location for a new buried utility within an existing, usually congested, right-of-way. In order to accomplish this, a decision support system for utility routing was developed as par of a much larger infrastructure knowledge portal being developed at the University of Toronto [11]. The portal aims at establishing a semantic interoperable hub for the exchange of information between different players to coordinate the design of infrastructure systems and construction plans through integrating product and process knowledge. The portal is based on the use of three base ontologies for products (IPD-Onto), processes and actors. For more information on the portal, the reader should refer to [11].

8.3.1 OWL-Geospatial Data Wrapping Module

The following section presents the Web Ontology Language (OWL) wrapper that was created to transform geospatial infrastructure product data into a format that is compliant with the ontology. The need for this wrapping mechanism stemmed from the following:

- ◼ There is a need to bridge the gap between current de facto data-centered representations of infrastructure products and the proposed knowledge-centered paradigm of the ontology.
- ◼ Municipalities and utility companies currently rely on some sort of geospatial data representation for their infrastructure data repositories. In order to start utilizing the proposed ontological model, there is a need to tap into these vast databases and use them as the first step in populating the ontology.
- ◼ These is a need to facilitate the use of the ontology as an interoperability enabler among the various utility sectors. The wrapper is able to embrace virtually any infrastructure product data schema and transform it into an ontology-compliant format.

The following sections describe the overall architecture of the wrapping system, along with a semantic concept recommendation algorithm that was proposed to semi-automate the process of concept matching.

8.3.1.1 Data Wrapping: Background

The process of data "wrapping," or the more generic process of *mediation* between heterogeneous data sources, was first introduced in the 1992 landmark article by Gio Weiderhold [12]. This paper presents mediators as modules that occupy an explicit middle layer between the user application and the data repository. The purpose of this middle layer is to essentially detach the application domain from the data resource. With the advent of the Internet and the explosion of information sources, this concept of data mediation propagated into what is now referred to as data wrapping.

Wrapping a data source means retrieving data from the source and translating it into a common integrated data representation [13]. Data wrapping is rooted within the domain of database information integration. Data wrapping modules are able to overcome the syntactic differences that exist between various schema. These schema-matching approaches have been used in the domains of bio-informatics [13], Hyper Text Markup Language (HTML)-based Web information [14], and manufacturing [15].

Other approaches that attempt to resolve both syntactic and semantic inconsistencies between data sources have relied on coupling data wrapping modules with ontologies. Guan et al. [16] developed an ontology-based data wrapper to mediate between various GML schemas. Their architecture relies on a wrapper that utilizes an ontology for spatial features to match heterogeneous GML schemas. Lee [17] utilized the dynamic evolving nature of ontologies to create an "intelligent" wrapping system that relies on incremental learning.

Some researchers have approached the problem of data model interoperability from the strict context of model matching desperate. In terms of matching, two generic methods have been developed in prior studies [18]: (1) a linguistic-based approach that finds matched elements using their names or descriptions (e.g., comments extracted from specifications), and (2) a constraint-based approach that considers the similarities of certain constraints, such as data types of an attribute, schema hierarchical structures, and relations between elements. As will be discussed in the following sections, a semantic (linguistic) approach was used for data wrapping in this application.

8.3.1.2 System Architecture

The wrapping system proposed in this research uses the ontology to resolve semantic discrepancies between models and string matching for the syntactic discrepancies. The system is composed of three main matching procedures: product matching, attribute matching, and attribute value matching. Product matching is the first (and simplest) procedure and is performed to identify the infrastructure product (water pipe, gas valve, etc). Since infrastructure product geospatial data is always stored as layers of distinct products, the system assumes that each data file contains only one type of product.

The second procedure, attribute matching, matches a set of geospatial field names (commonly referred to as attributes in GIS terminology) to their equivalent infrastructure product attributes in the ontology. The third and final procedure, attribute value matching, matches a set of geospatial field values (commonly referred to as attribute values in GIS terminology) to their equivalent infrastructure product attribute domain lists in the ontology. For example, an attribute could be "Gas Pipe Material" while its attribute value could be "High-Density Polyethylene."

The wrapping module utilizes three inputs. Two of these inputs are necessary, while the third is optional. The two necessary inputs are the ontology (in OWL format) and the infrastructure geospatial data file (in XML format). The optional input is the standard matching schema (in XML format).

The output of the wrapping module is an ontology-compliant OWL instance file. Optionally, the module can save the user's matching results to a user-defined matching schema that can be reused.

Figure 8.4 shows the inputs and outputs to the wrapping system. Items indicated with a dotted arrow are optional, while those indicated with a solid arrow are necessary.

Reusing standard and user-defined matching schema is useful in accelerating the wrapping process. This feature is intuitive due to the relatively static nature of data standards. By creating a set of matching schemas between the ontology

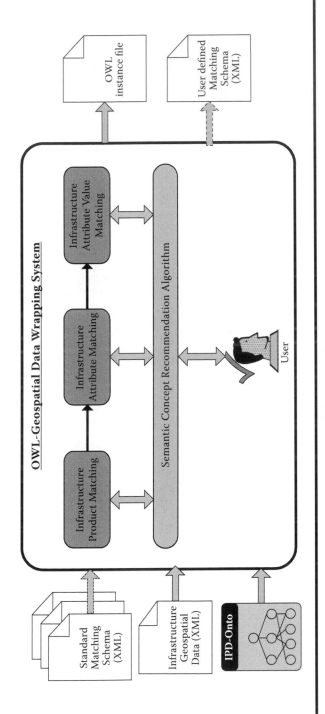

Figure 8.4 System architecture for OWL-Geospatial data wrapping system.

and the commonly used infrastructure product data standards (e.g., SDSFIE [19], MultiSpeak [20], vendor data models), the wrapping system can quickly convert geospatial data into ontology-compliant OWL files without human interaction. The system also allows the creation of user-defined matching schemas to be reused, hence eliminating the need for repetitive user input during the matching process.

With around 1,800 concepts in the ontology, manually performing this matching would be a daunting task. As such, a semantic concept recommendation algorithm is proposed to present a short-list of similar concepts for the user to confirm. Details of this algorithm are discussed in the following section. This concept recommendation is performed at the later two levels of matching (attribute and attribute value).

8.3.1.3 Semantic Concept Recommendation

The process of wrapping the infrastructure product geospatial data into an ontology-compliant OWL instance file is a semi-automatic process. The wrapper uses a semantic recommendation algorithm to recommend a concept from the ontology as a potential matching candidate to the geospatial concept. The user is presented with a set of potential concepts from the ontology (along with their definitions). The user then selects the concept from the ontology that is a best fit (equivalent or very similar) to the geospatial concept. The user must always confirm any concept matching (hence the semi-automatic process).

The matching procedure utilizes both semantic and syntactic matching algorithms as shown in Figure 8.4.

8.3.1.3.1 Semantic Matching

The recommendation procedure follows four stages of semantic concept matching. Semantic matching can be performed at the element-level or the structure-level [17]. Element-level semantic techniques analyze individual labels or concepts at nodes, while structure-level techniques utilize the hierarchy and relationships of the ontology in matching. The first stage of the procedure can be considered an element-level approach, while the other three stages are structure-level approaches.

The four stages of semantic matching proceed in decreasing order of semantic similarity as depicted in Figure 8.5. Concept recommendation proceeds in the following order:

1. Field-to-Concept Matching: This approach matches an individual geospatial attribute (or attribute value) to the actual name of the OWL class in the ontology. Matches are made to the *owl:class* tag.
2. Equivalent Concept Matching: This approach relies on the built-in OWL class *owl:equivalentclass* to match an individual geospatial attribute (or attribute value) to a concept that is equivalent. Matches are made to the *owl:equivalentclass* tag.

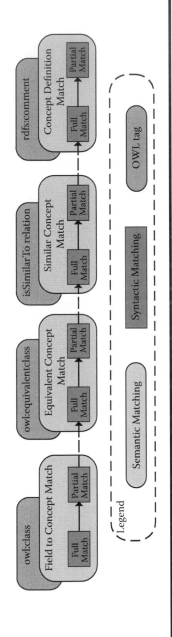

Figure 8.5 Procedure for semantic recommendation algorithm.

3. Similar Concept Matching: This approach utilizes the *isSimilarTo* relationship that was discussed in Section 8.0. Matches are made to the *isSimilarTo* tag.
4. Concept Definition Matching: This approach relies on the fact that all concepts in the ontology have a natural language definition assigned using the *rdfs:comment* tag. Matching is performed between the geospatial attribute (or attribute value) and the text definition in the *rdfs:comment* tag.

For example consider the following OWL abstract syntax description for a Gas Vault:

```
Class(GasVault complete annotation(rdfs:comment "Vaults
protect and provide access to underground gas pipes and
valves. ") GasJunctionBox)
SubClassOf(GasVault p1:GasProduct)
SubClassOf(GasVault restriction
        (p1:isSimilarTo
              someValuesFrom(GasStationStructure)))
```

The semantic recommendation algorithm will first attempt to match the geospatial concept to the OWL class name: "*GasVault*." If no match is found, the algorithm will attempt to match to the equivalent concept of "*GasJunctionBox*." If still no mach can be found, the algorithm will attempt to match to the *isSimilarTo* relationship of "*GasStationStructure*." Finally if no match is found, the recommendation algorithm will search in the definition "*Vaults protect and provide access to underground gas pipes and valves.*"

8.3.1.3.2 String Matching

The string matching problem involves the process of finding the occurrence of a pattern of text within a larger body of text. String matching is sometimes referred to as *syntactic* matching as there is no comprehension of the meaning of the text being searched. Cormen et al. [22] formalize the string-matching problem as follows:

■ Assume that the text is an array $T[1...n]$ of length n.
■ Assume that the pattern is an array $P[1...m]$ of length $m \leq n$.
■ Assume the elements of P and T are characters drawn from a finite alphabet Σ.

We say that pattern P occurs with shift s in text T (or, equivalently, that pattern P occurs beginning at position $s+1$ in text T) if $0 \leq s \leq n-m$ and $T[s+1...s+m] = P[1...m]$. Several algorithms have been developed to address the string matching problem (Naïve, Rabin-Karp, Finite Automation, etc.). The wrapper implements the Naïve (brute-force) algorithm. The naive algorithm finds all valid shifts s using a loop that checks the condition $P[1...m] = T[s+1...s+m]$ for each of the $n-m+1$ possible values of s.

For each of the aforementioned four recommendation procedures, a text search algorithm performs Full Matching followed by Partial Matching.

8.3.1.3.2.1 Full Matching — A full match occurs when the attribute or attribute value defined in the geospatial data schema exactly matches a concept from the ontology. In the algorithm terminology, this assumes that m = n, or the length of the text must equal the length of the pattern.

For example, the following is an example of a full match:

"Length" = "length" or "Cross Section" = "Cross_Section"

8.3.1.3.2.2 Partial Matching — A partial match occurs when the attribute or attribute value defined in the geospatial data schema partially matches a concept from the ontology. Because spaces are not allowed between words using the OWL syntax, the ontology relies on underscores or capitalization to indicate a break between words (e.g., PipeLength or Electricity_Switch). The recommendation algorithm utilizes this fact to prevent incorrect string matching. In partial matching, each of the constituent words of the ontology concept is compared to the geospatial data field. For example, the following is an example of a partial match:

"MaterialType" = "PipeMaterial" or "Valve" = "Water_valve_type"

8.3.1.4 System Features and Issues

This section provides a summative assessment of the OWL-geospatial data wrapping system by highlighting its key strengths and weaknesses. Some of the useful features included in the system's requirement include:

■ *Flexibility to changes in the ontology.* The wrapper was designed to take into account the possibility of changes to the ontology (e.g., adding attributes, re-classification of attributes, changes to attribute domain lists). The flexibility is limited in the sense that it will not accommodate fundamental changes to the upper ontology (e.g., removing the concept of attributes or products altogether). Changes of this kind are unlikely as they would defy the purpose of the ontology.
■ *Ability to create, load, and edit existing matching schemas.* This feature significantly decreases the required user effort in creating matching schemas. This is primarily due to the fact that (1) several matching schemas between the ontology and standard infrastructure product data models were created, and (2) user-defined matching schemas can be easily updated to reflect changes in their structure.
■ *Ability to match only those attributes or attribute values that are actually being used in a geospatial data file.* This feature is specifically useful in cases where partial models are populated with data. From a practical perspective this is very often the case due to the broad coverage of data models and the scarcity of reliable data related to infrastructure.

8.4 Case Study: Infrastructure Routing

The following case study illustrates how information from CAD and GIS sources was marshaled based on IPD-Onto using the semantic data wrapper to support the process of urban infrastructure routing. The project involved the replacement of a 1,050 m distribution water main along a major thoroughfare in the City of Toronto, Canada. The existing 200 mm cast iron water main was being replaced by a 400 mm PVC pipe on Walsh Avenue from Weston Road to Wilson Avenue, and a 300 mm PVC pipe on Weston Road from Blondin Avenue to Starview Lane. The project was completed in 2005 and cost the City of Toronto $700,000.

The project was selected because of its relatively accurate and detailed level of buried utility information. As a pilot project, the North York District of the City of Toronto conducted a subsurface utility engineering (SUE) investigation. Reliability and accuracy of utility information is vital for conducting a micro-level routing analysis. The SUE investigation that was conducted provided accurate locations for existing main structures (pipes, lines, etc.) for all buried utility systems. The investigation did not locate supporting structures or devices (valves, chambers, manholes, etc.) for non-municipal utilities. The city had existing records of its support structures and devices. The resulting composite utility map (Figure 8.6) consisted of:

- Main structures for all utilities (electricity, gas, telecommunication, water, and wastewater)
- Fire hydrants, valves, and junctions for all water utilities
- Catch basins and manholes for stormwater infrastructure
- Manholes for sanitary sewer infrastructure

In addition, the following street features were included in the routing analysis:

- Trees
- Driveways for residential units
- Surrounding land use

8.4.1 Data Processing

The raw data, which was obtained from the City of Toronto was processed using the semantic data wrapper. This processing involved three main steps:

1. CAD/GIS data integration: The data provided by the city was in both CAD and GIS formats. The consultant that performed the SUE study delivered the non-municipal utility mapping in CAD; the design of the new water main was in CAD; but the city's existing utility assets and land use were in GIS. All CAD data was transferred to GIS. The resulting composite utility and land use data was transformed into the ESRI Geodatabase format (Figure 8.7). This task took one-and-a-half hours.

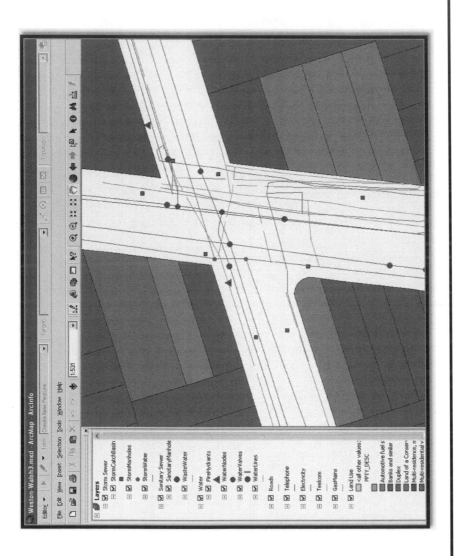

Figure 8.6 Composite utility map and land use for the City of Toronto case study.

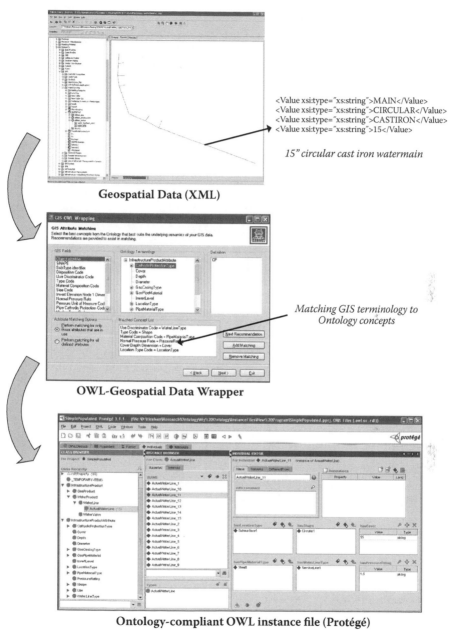

<Value xsi:type="xs:string">MAIN</Value>
<Value xsi:type="xs:string">CIRCULAR</Value>
<Value xsi:type="xs:string">CASTIRON</Value>
<Value xsi:type="xs:string">15</Value>

15" circular cast iron watermain

Geospatial Data (XML)

Matching GIS terminology to Ontology concepts

OWL-Geospatial Data Wrapper

Ontology-compliant OWL instance file (Protégé)

Figure 8.7   Screen shots of OWL-geospatial wrapper.

2. Wrapping: This data was subsequently transformed into an ontology-compliant format using the OWL geospatial data wrapper that was described in the previous section. Due to the fact that the city's data did not follow any standardized schema, the matching process between the city's data model and the ontology took two-and-a-half hours. Had a standard matching schema been available, the wrapping process could have been completed in approximately 15–20 minutes. The wrapping procedure is depicted in Figure 8.6.

3. Constraint Reasoning: The third and final step involved using the OWL Spatial Constraint Satisfaction Reasoner that was described in Section 8.0. The system allows the user to select:
   - *Product Options:* Specifies which products in the ontology to check for.
   - *Constraint Options:* Specifies which subset of constraints to check for (by constraint type and constraint purpose).
   - *Proximity Threshold:* Used for Location Preference constraints that are not qualified by a specific distance. This allows the user to specify a distance threshold that triggers the constraint.
   - *Temporal Specification:* Used if the user wants to check for constraints that are valid within a specific time range (e.g., seasonal constraints, road closure constraints, intermittent business requirement constraints).

8.4.2 Analysis Results

Using the micro-level routing system to evaluate the route of the proposed water main revealed the following:

■ No clearance constraints were found to be violated, only location preference constraints (Figure 8.8). This is due to the fact that the system analyzed the final design of the water main after the necessary revision cycles (both internal within the City of Toronto and external with impacted utilities). It would be interesting to perform the analysis on an earlier stage of design (e.g., 30% design stage) in order to evaluate the value of the clearance constraints within the knowledge base.

■ Four location-preference constraints were found to be violated at a 2 m distance threshold. In total, 26 pairs of products were violating these location-preference constraints. A listing of constraint-violating products at various distance thresholds is given in Table 8.1

■ It is noted that at increasing levels of distance thresholds, the number of constraint-violating product pairs rises significantly. This is due to the constraint losing "context." For example, a new utility line that is below a manhole but at a horizontal distance of 7 m will most likely not be impacted by

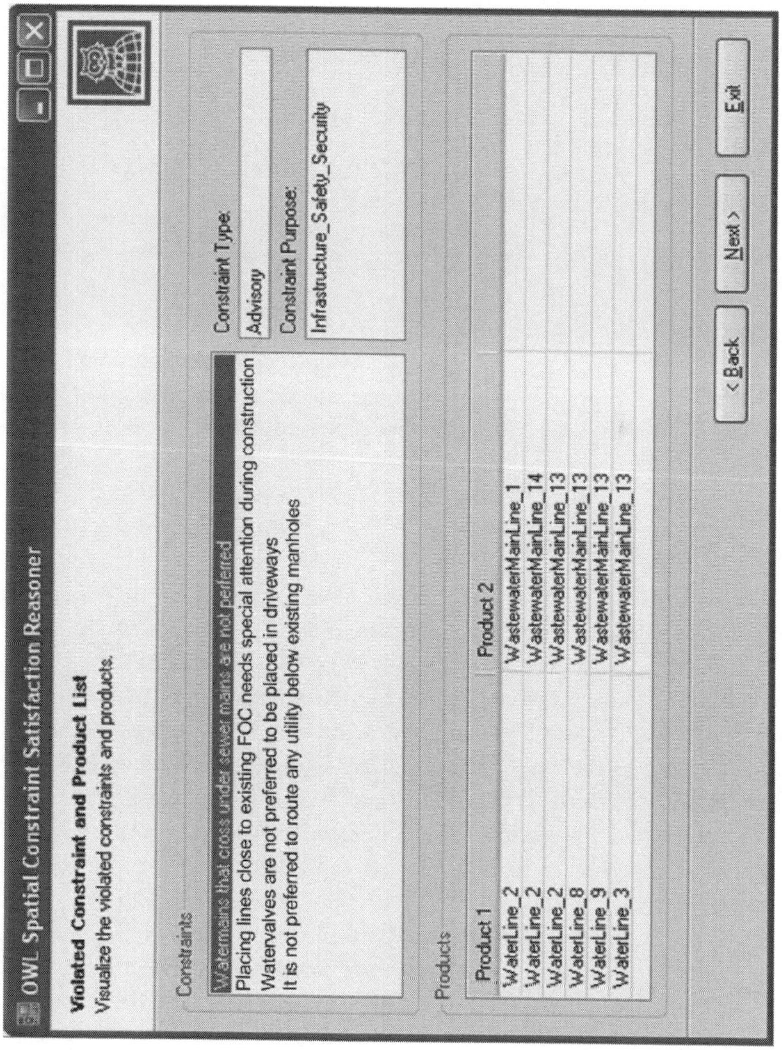

Figure 8.8   Constraint result screen within the OWL Spatial Constraint Satisfaction Reasoner.

Table 8.1 Constraint-Violating Products at Various Distance Thresholds

Constraint	Threshold (m)			
	0.5	2	4	8
Watermains crossing under sewer mains.	2	6	14	33
Placing new lines close to fiber optic cables.	3	9	11	25
Water valves along driveways.	4	4	4	4
New utilities below manholes.	4	7	10	27
Total number of product pairs violating constraints	13	26	39	89

the manhole. As such, the thresholds should be used by the routing engineer within the project's context and not absolutely. One constraint could be relevant at a 6 m threshold, while another constraint could be relevant to only 1 m. This "relevancy" is highly case specific and hence is not directly represented within the ontology.

8.4.3 System Evaluation

Based on feedback from City of Toronto engineers involved in the project, the following potential benefits would have been attained from using the system:

1. Spatial data integration time savings: An extensive amount of time is spent by city technicians in assembling and integrating data from geospatial and non-geospatial sources. The data integration capabilities provided by the ontology and made possible through the use of the OWL wrapper would have eliminated an estimated 40–60 hours of data preparation by city staff.
2. Automated constraint checking time savings: City staff were specifically interested in the clearance constraints used by the system based on the Toronto Public Utility Coordination Committee (TPUCC) requirements. The fact that the GIS system contained information on horizontal and vertical locations of buried infrastructure allowed a quick (and accurate) check to be performed against TPUCC requirements. Traditionally these checks were time-consuming (and sometimes inaccurate) due to the fragmented nature of the data (CAD files, specifications, and survey information).
3. Improved utility coordination meeting efficiency: City staff usually conduct several utility coordination meetings with utility companies that operate infrastructure within the vicinity of any proposed project. On this specific project

two meetings were held. The use of the Web-based GIS system would have greatly improved the efficiency of these meetings by allowing most technical issues to be exposed.

4. Utility review and approval time savings: Approval of a proposed route from utility companies can sometimes take a considerable amount of time. This is due to (1) the traditional hard copy drawing circulation procedure required by some utility companies, and (2) the manual constraint checking process used in the review. On this project, utility review and approval consumed 9 business days (until all approvals were received). The use of the Web-based GIS system would have significantly reduced this time (if users are relatively prompt it should not take more than one business day).

8.5 Summary

Although originally intended as a mechanism to represent and share domain knowledge, interest in ontologies has rebounded as an enabler of interoperability. With their rich semantic modeling capabilities, ontologies offer an upper-hand over data models with their ability to address knowledge interoperability. As such, this chapter presented a succinct overview of ontologies with a detailed description of an ontology that was developed to represent utility infrastructure products and bridge the interoperability gap that commonly exists between infrastructure agencies in urban settings. A use-case scenario surrounding utility route selection was developed, and an application ontology was created to address the use case. As an interoperability enabler, the ontology was capable of fusing several data models in both CAD and GIS into an ontology-compliant format. This mediation was accomplished by a semantic wrapping tool built around the ontology. The tool was successful in streamlining heterogeneous data from various utility agencies for the purpose of optimized infrastructure route selection. Using an ontology coupled with a semantic wrapper as a mediator shows promise in both bridging the worlds of CAD and GIS and achieving inter-organization interoperability.

References

Antoniou, G. & Van Harmelen, F. *A Semantic Web Primer*. MIT Press. Cambridge, MA, 2004.

Cormen, T., Lieserson, C, Rivest, R., and Stein, C. (2001). *Introduction to algorithms*. Boston, MA: MIT Press.

El-Diraby, T.E. (2006) "Infrastructure development in the knowledge city" Lecture Notes in Computer Science, Intelligent Computing in Engineering and Architecture—13th EG-ICE Workshop, ASCONA, Switzerland, 2006, p. 175–185.

El-Diraby, T.E., Lima, C., and Fies, B. (2005). "A core taxonomy for construction systems: Toward semantic exchange of project knowledge," *J. of Computing in Civil Engineering*, ASCE, 19 (4).

Fensel, D. *Ontologies: A Silver Bullet for Knowledge Management and Electronic Commerce,* Springer-Verlag, 2001.

Gallaher, M., O'Connor, A., Dettbarn, J., and Gilday, L. (2004) Cost Analysis of Inadequate Interoperability in the U.S. Capital Facilities Industry. National Institute of Standards and Technology.

Giunchiglia, F., and Shvaiko, P. (2004). "Semantic matching." *The Knowledge Engineering Review.* 18(3): 265–280.

Gómez-Pérez, A , Lopez, F.P. and Corcho, O. *Ontological engineering: with examples from the areas of knowledge management, e-commerce and the semantic Web,* Springer-Verlag, 2005.

Guan J., Zhou, H., Chen, J., Chen, X., An, Y., Yu, W., Wang, R., Liu, X. (2003) Ontology-based GML schema matching for spatial information integration International Conference on Machine Learning and Cybernetics. 4: 2240–2245.

Lacroix, Z. (2002). "Biological data integration: Wrapping data and tools*" IEEE Transactions on Information Technology in Biomedicine,* 6(2): 123–128.

Lake, R., Burggraf, D, Trininic, M., and Rae, L. (2004) *Geography Markup Language: Foundation for the Geoweb.* New York: John Wiley.

Lee, S. (2004). "XML schema matching based on incremental ontology update" Web Information Systems—WISE 2004. 5th International Conference on Web Information Systems Engineering. Proceedings (Lecture Notes in Computer Science Vol. 3306), 2004, 608–618. Berlin, Germany: Springer-Verlag.

Liangyu, Z. and Nagi, R. (2002) "Design of distributed information systems for agile manufacturing virtual enterprises using CORBA and STEP standards" *Journal of Manufacturing Systems,* 21(1): 14–31.

Maciaszek, L. (2005). *Requirements analysis and system design* (2nd ed.). Harlow, U.K.: Addison Wesley.

May, W., and Lausen, G. (2004) "A uniform framework for integration of information from the Web" *Information Systems,* 29(1): 59–91.

Missikoff, M. A tourism ontology for small and medium enterprises in European market. LEKS, FETISH Project, Deliverable D1.1, IASI-CNR, Rome, 2000.

MultiSpeak. http://www.multispeak.org/whatisit.php. Accessed July 2005.

Rahm, E., P.A. Bernstein (2001). "A survey of approaches to automatic schema matching" *VLDB Journal.* Vol. 10 334–350.

Spatial Data Standard for facilities, infrastructure, and environment—Data model and structure, U.S. Army CADD/GIS Technology Center, 2007.

Spyns, P., Meersman, R. and Jarar, M. Data Modeling versus Ontology engineering. *ACM SIGMOD Record,* 31(4): 12–17, 2002.

Unified Medical Language System http://www.nlm.nih.gov/research/umls/. Accessed July, 2005.

Wiederhold, G. (1992). "Mediators in the architecture of future information systems." *Computer* 25(3): 38–49.

Chapter 9

CAD and GIS Interoperability through Semantic Web Services[*]

Burcu Akinci, Hassan Karimi, Anu Pradhan, Cheng-Chien Wu, and Greg Fichtl

Contents

---

[*] This material was originally published in *ITCon*, Vol. 13, pp 39–55. Available at http:/www.itcon.org/2008/3

9.1 Introduction

Civil engineering projects are multi-disciplinary in nature involving a large number of participants, such as designers, engineers, project managers, and construction managers. Since the advent of computer-aided design (CAD) and geographical information system (GIS) tools, project participants have been increasingly leveraging these tools throughout the different phases of a civil infrastructure project. For instance, during construction of a facility in a densely populated area, a CAD system augmented with construction schedule information (also known as 4D CAD) would be employed to detect spatio-temporal conflicts between a crane and concurrent construction activities at a job-site, and a GIS would be used to plan an optimal route that minimizes traffic congestion for delivery of construction materials to the site.

Civil engineering tasks require that CAD and GIS platforms be interoperable as data or analyses results generated by one system (CAD or GIS) are often required by the other. For example, a set of spatio-temporal conflict results produced by a CAD system can be used by GIS to calculate possible time frames for delivery of construction materials to avoid further spatial conflicts. However, as existing CAD and GIS platforms have been developed independently with different purposes, there are significant differences in terms of data formats they support, terminology they utilize, semantics of concepts they represent, and reasoning techniques on which they are based. Participants of civil infrastructure projects access both CAD and GIS during different stages of a project to perform different tasks. Often the completion of an engineering task requires translation of information created or maintained in one system (CAD or GIS) for use by the other system (Jones 2005). Existing solutions, commercial and non-commercial, to the interoperability problem have focused on developing data exchange formats between CAD and GIS platforms. For instance, major CAD and GIS software packages provide data exchange between these platforms (e.g., see Autodesk 2007; ESRI 2007).

Realizing interoperability as being an important issue within their respective domains, a variety of Architecture, Engineering and Construction (AEC) and geospatial consortiums have focused on developing standards to enable seamless data transfer and interoperability among software systems within each domain. For example, International Alliance for Interoperability (IAI) is specifying data standards, such as Industry Foundation Classes (IFC), for the architecture, engineering, construction, and facility management (AEC) community (IAI 2007), and Open Geospatial Consortium (OGC) has been carrying out similar standardization efforts, such as Geography Markup Language (GML), for the geospatial community (OGC, 2006). Recently, the interest in inter-domain interoperability between AEC and geospatial domains has spurred new standardization efforts such as IFC 2x3G specification (IAI 2007). While these efforts have focused on enabling data exchange between various CAD and GIS platforms, they have not addressed issues related to differences in semantics and reasoning capabilities between them. This is the reason why, to achieve full interoperability, there is a need for semantic interoperability solutions and reasoning techniques between CAD and GIS platforms. For example, some spatial analysis functionalities (e.g., buffer and spatial query) available in GIS are not available in CAD (Rasdorf et al. 2000). Similarly, CAD systems can perform operations (e.g., spatial conflict detection) at a finer level of detail compared with GIS due to differences in the spatial scale of the objects represented. Where CAD and GIS platforms are unable to resolve their semantics issues and to realize their reasoning capabilities, they would not be fully interoperable, resulting in processing of many time-consuming tasks manually with a high level of ambiguity.

In this chapter, we present a potential approach toward bridging the interoperability gap between CAD and GIS platforms. The premise of this approach are ontologies to address semantic differences between the AEC and geospatial domains, and Web services, and to allow dynamic composition of CAD and GIS operations needed to complete a specific task. We begin with a motivating scenario that is focused on the management of equipment space requirements to highlight the need for interoperability between the AEC and geospatial domains. We describe the components of the proposed semantic Web service approach: task decomposition, ontology identification, Web service discovery and matching, and service composition. We conclude by highlighting research challenges associated with implementation of some of these components.

9.2 Example Scenario

Construction projects are becoming more complex as space on construction sites get tighter and more construction activities are scheduled concurrently. In such cases, space management required by various types of equipment becomes increasingly challenging (Akinci et al. 2003; Tantisevi and Akinci 2007). Ineffective space management results in conflicts, which can create work interruptions, productivity reductions, hazardous work conditions, and damage to existing structures (Guo 2002; Varghese and O'Connor 1995). An example of an engineering task that involves space management is crane location analysis for construction sites. To ensure that a crane is safely located, all possible spatial interactions between a crane and existing structures on and around a job site need to be analyzed. These existing structures may include objects within a construction site, such as portions of facilities being built, existing equipment, material staging locations, and subsurface utilities. Similarly, possible spatial interferences between objects with close proximity to the job site, such as nearby buildings and power lines, should be analyzed. Finally, when a crane is located on an existing roadway, possible traffic impacts of lane closures due to crane operation must also be considered.

While some of these analyses, such as identification of possible spatial conflicts between a crane and the facility that is under construction, can be performed by using data obtained from CAD and construction schedules; others, such as analysis

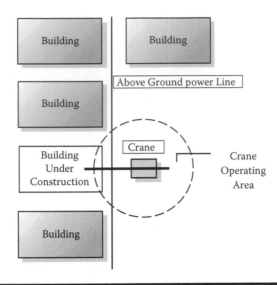

Figure 9.1 Potential obstructions for a crane based on its workspace.

of impacts of possible crane locations to subsurface utilities or nearby structures and power lines, need to be performed using data from both CAD and GIS. For the purpose of this paper, we will focus on the analysis related to identification of potential obstructions that are close to a possible crane location, such as neighboring buildings or surface utilities, as an example to highlight interoperability challenges associated with CAD and GIS integration. Determining a suitable location for a crane involves analyzing geometric constraints based on crane specifications and the dimensions of the building under construction, analyzing the load of the component to be lifted, and determining the feasible area to pickup the load based on both geometric and load limit calculations. These analyses, however, will only ensure a safe operating area for a given crane with respect to the building under construction. In a real world job site, it is likely that there are multiple possible obstructions including neighboring buildings and other objects, such as above-ground power lines (Figure 9.1).

Detecting possible interferences between existing structures and the crane's workspace requires completion of a set of sub-tasks including: (1) identifying the "candidate" (i.e., potential) obstructions; (2) gathering, transforming, and assembling data needed for interference detection; (3) determining true obstructions from the candidates list; and (4) preparing the results for review. There are several approaches to accomplish the abovementioned sub-tasks. One approach is to create 2D drawings or sketches of a job site and overlaying the proposed position of the crane. This is a manual process and, while it may be adequate, it may not include all potential obstructions or analyze possible interactions from a 3D or temporal point of view (Tantisevi and Akinci 2007). To overcome such shortcomings, the second approach is to model the job site, its surroundings, and the crane in 3D and perform clash detection to identify possible spatial conflict. Both (2D and 3D) approaches leverage data and operations performed in CAD and GIS environments; albeit the nature of the data and the operations utilized are different based on whether a 2D or 3D analysis is being performed. Figure 9.2 depicts these two approaches (note that there are other alternative approaches) to complete the task of identifying possible obstacles associated with a given crane location.

Both approaches (depicted in Figure 9.2) perform the required sub-tasks of identifying candidate obstructions, transforming and assembling data to identify possible spatial conflicts, performing conflict analysis, and preparing and highlighting

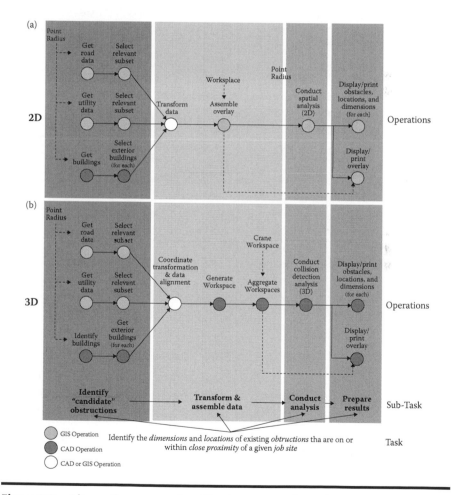

Figure 9.2   Alternative approaches 2D (a) versus 3D (b) analyses for analyzing possible obstacles that might interfere with a crane operation during construction.

the resultant obstacles. However, each approach performs the required analyses using a different combination of CAD and GIS operations. For example, when analyzing potential obstructions, the 2D approach is more GIS-centric and uses a 2D spatial analysis operation (Figure 9.2a), while the 3D approach is more CAD-centric and performs a 3D collision detection analysis (Figure 9.2b). While similar categories of data are needed (e.g., neighboring buildings, utilities, roads), the level of detail may differ based on the requirements of the selected operations. For instance, in 2D analysis, the heights of power lines and buildings are not available, while the height and detailed geometric information in 3D space are needed for 3D analysis. In addition, the direction of data transfer from one platform to another and corresponding data transformation are different in each of these approaches. For example, in the 2D approach, the local coordinate system used by CAD must be manually matched to the global coordinate system used by the GIS; on the other hand, in the 3D approach, the global coordinate system of existing structures within the GIS need to be transformed into the local referencing system used within the CAD.

Given the range of CAD and GIS operations needed to perform such analyses and the lack of interoperability between CAD and GIS, currently it is necessary for engineers to have both CAD and GIS skills to be able to perform engineering tasks that require data and operations between both platforms. In addition, currently both AEC and geospatial domains are faced with semantic ambiguities. For example, consider the term "obstruction." What constitutes an obstruction in the AEC domain may be markedly different from the same concept in the geospatial domain. A GIS expert may apply a definition inconsistent with the AEC domain's definition, and thereby fail to properly evaluate all potential obstructions. As we have seen with this scenario, while the approaches described in Figure 9.2 can reduce the risk associated with the placement of a crane on a construction site, there are additional challenges that may make this analysis too difficult and/or too costly to perform. Within the context of this chapter, we suggest a solution approach focused on semantic Web services. Our vision for this approach will be discussed in the next section.

9.3 Vision: Semantic CAD/GIS Web Services

The Semantic Web offers a common framework that allows data to be shared and used across multiple applications and communities (W3C 2007). It is a collaborative effort initiated by the World Wide Web Consortium (W3C) with participation from a large number of researchers, academic institutions, and industrial partners. The Semantic Web leverages ontologies and standard languages, such as Resource Definition Languages (RDF) and Web Ontology Language (OWL) for recording machine-readable data and defining ontologies, respectively. In our proposed approach, we consider the Semantic Web as a common framework for interoperating CAD/GIS operations. Figure 9.3 depicts the overview of our envisioned

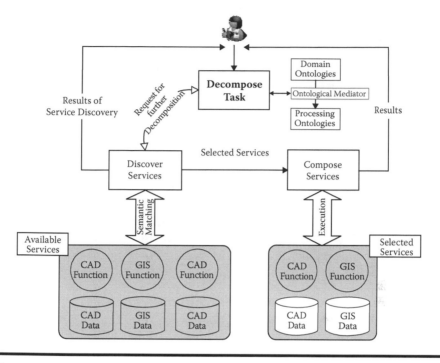

Figure 9.3   Semantic CAD/GIS Web services vision.

approach of using semantic Web services for achieving interoperability between CAD and GIS. The approach consists of three modules: (1) task interpretation, (2) Web-service matching, and (3) Web-service composition.

The vision is task-oriented and begins with a user defining a specific geospatial analysis task (Figure 9.3). Using our scenario as an example, a task would be "identify the dimensions and locations of existing obstructions that are on or within proximity to a job site." Additional information must be provided along with needed parameters and constraints depending on the context of the analysis that must be performed. For example, information on the model of the crane to be utilized, the building model for the building under construction, and site related data may be needed.

In this vision (Figure 9.3), AEC domain ontologies are proposed to be used to decompose a given task into a set of sub-tasks (using the *Decompose Task* module), which in turn, through the *Ontological Mediator* they are mapped to specific CAD or GIS operations (Peachavanish and Karimi 2007). The outcome will be a workflow of operations and data transfers that will be used to discover available CAD/GIS Web services and to compose them in answering the given task. Using our scenario example, the workflow generated after the decomposition task would include operations to gather and transform data (e.g., from a CAD-based building information model to a vector-based GIS model), to assemble an overlay aligning the various elements under consideration (e.g., crane, building under construction, adjacent buildings, surface utilities) with the underlying geospatial coordinate reference system, and to conduct spatial analysis to identify obstructions.

Once the required operations are identified, the search for available services can begin with the help of the *Discover Services* module. For each operation, a matching algorithm in the *Discover Services* module evaluates existing services against the requested service (i.e., an operation). If a suitable service cannot be identified (i.e., no match is made), the operation will be further decomposed into lower level operations for which there may be match services. Further, in *Discover Services,* each identified service will be evaluated based on Quality of Service (QoS) parameters. In the event that the available service for a specific operation does not meet the QoS parameters, feedback will be provided to the user and they will be given an opportunity to accept or reject the service. If no service can be located, the task cannot be completed and feedback will be provided to the user. If exceptions are encountered during execution (e.g., a service has become unavailable in the interim between identification and execution), feedback is provided to the user.

When matches between CAD/GIS Web services for a given task are found, they need to be chained together (i.e., composed) and invoked in a specific order to provide the requested outcome. Service composition is done with the help of the *Compose Services* module. In a highly dynamic environment like Web services, service composition is susceptible to a multitude of sources of uncertainties, including network latency, availability of services, and quality of available services. A planning-based approach that can handle uncertainties of Web service composition is the final module in our approach.

9.4 Components of the Semantic CAD/GIS Web Services Approach

To take full advantage of Web services (allowing users to assemble operations based on the needs of each specific project), which are expected to be numerous for each domain, they need to be searched and matched semantically. To semantically search and match Web services, ontologies, which include ontologies that define specific concepts within a domain and those general ontologies that define relevant concepts to the task at hand, are needed. Our proposed semantic Web services approach for CAD/GIS integration requires that CAD/GIS ontologies be used to resolve potential semantics issues in deciding appropriate Web services for CAD/GIS operations. The details of what these CAD/GIS ontologies should be, how they could be used for CAD/GIS problem solving, what CAD/GIS Web services should be, and how they could be used for CAD/GIS integration are given in this section.

9.4.1 Ontologies

The key to our proposed approach is a set of ontologies, primarily domain ontologies, that upon submission of any given task help resolve semantic issues associated with CAD/GIS integration. Ontology is defined as an "explicit specification of a conceptualization" (Gruber 1993). Generally, it is represented as a set of concepts within a domain and the relationships between the concepts. The specification of an ontology comprises a vocabulary of terms where each term defines its meaning (Boury-Brisset 2003). Ontology has been used in various areas such as knowledge management (Fensel 2002), semantic Web (Fensel et al. 2001), and data fusion (Boury-Brisset 2003).

Ontologies are becoming an increasingly important research area in the field of geospatial information science. Recent research in the area of geospatial ontology has been focused on the formal modeling of the geospatial world (Mark et al. 1999; Smith and Mark 2001), allowing for cross-system interoperability (Karimi et al. 2003; Peachavanish et al. 2006), geospatial data integration ((Cruz et al. 2004; Fonseca et al. 2003; Fonseca et al. 2002), and facilitation of geospatial information retrieval in heterogeneous networked environments (Klien et al. 2006). AEC domain ontologies (e.g., IFC and Barbie) define concepts, activities, and objects, and the relationships among elements defined within the AEC/CAD domain. In 2006, OGC examined the feasibility of representing GML in OWL as part of a preliminary effort to extend existing services, encodings, and architectures with Semantic Web technologies (OGC 2007). Further, the Geospatial Incubator Group of the W3C has focused on addressing issues of location and geographic properties of the Web of today and tomorrow (W3C 2007). This group recognizes ontologies as a critical part of its scope and the development of recommendations for geospatial ontologies as a key short term objective.

Within the geospatial domain, multiple ontologies based on ISO, OGC, and Federal Geographic Data Committee (FGDC) standards have been developed and made available on the Internet. In the AEC domain, research efforts investigate the opportunities to leverage the current IFC model to derive ontologies and develop standard models of the knowledge within the construction domain (BARBi 2007; e-Cognos 2007; El-Diraby et al. 2005; El-Diraby and Kashif 2005). While both communities (geospatial and AEC) have been actively engaged in developing their own sets of standards to enable interoperability among different software systems within their respective domains, only recently has the aspect of interoperability among *inter*-domains been officially recognized.

9.4.1.1 Ontologies for CAD/GIS Problem Solving

As discussed in the scenario in Section 9.2, there may be multiple approaches to complete the needed analysis based on the characteristics of the task and the desired level of accuracy. The specific approach to conduct the analysis will drive a set of operations that need to be executed to provide the desired outcome of the given task. For example, for a site in a densely populated urban area, a more detailed 3D analysis may be needed than for a site located in a rural area, where a 2D spatial analysis may be sufficient.

In addition to utilizing ontologies for Web service matching (discussed later in this paper), ontologies in CAD/GIS integration could be used to understand and interpret engineering tasks. This can be accomplished by decomposing a given task into its individual sub-tasks and identifying the required CAD and GIS operations that must be performed. Such a decomposition of a task requires ontologies in both AEC and geospatial domains. For example in our scenario, the task, "identify the dimensions and locations of existing obstructions that are on or within proximity to a job site," could be decomposed into the following sub-tasks: *identify candidate obstructions, transform and assemble data, conduct analysis,* and *prepare results.* Clearly, performing these four sub-tasks requires both CAD and GIS operations, ranging from data transformation to spatial analysis to various computations to map generation.

Once a task is decomposed into a set of sub-tasks, each sub-task will be matched with a specific operation or a set of operations defined in CAD/GIS processing ontologies using the AEC-CAD/GIS Mediator that defines the relationship between domain level sub-tasks and processing operations. Each operation is defined in the processing ontologies with an identifier, one or more input parameters, and one or more output parameters. For example, the operation TRANSFORM takes a data file and a target format as input and produces a new data file in the new format as an output. The processing ontologies must support the definition of both basic operations, which cannot be broken down into simpler operations, and compound operations, which are operations made up of two or more basic (or other compound) operations. Once the operations are identified along with their parameters, a workflow will be constructed. An alternative approach to identify the sub-tasks of a task and construct a workflow is to use Natural Language Processing (NLP) to interpret tasks. In this case, the task must be presented in a language with a specific structure common in the domain. NLP has been the subject of significant research to automate the creation of Conceptual Data Models (CDM) such as Entity Relationship Diagrams (ERD) from requirements specifications (Ambriola and Gervasi 2006; Chen 1997; Harmain and Gaizauskas 2000; Mich 1996).

9.4.1.2 Identification, Evaluation, and Selection of Ontologies

As the use of ontologies to establish formal semantic agreements gains popularity, it is unrealistic to expect that there will be agreement on a single ontology or even a small set of ontologies within each domain. It is more likely that a vast number of ontologies within and across domains will be developed. Given the vast number of ontologies within and across domains, identifying relevant ontologies and reconciling among different ontologies is critical and has resulted in approaches to map (i.e., establish links between ontologies) or merge (i.e., generate a unique ontology from a set of original ontologies) multiple ontologies (Kotis et al. 2006; Noy and Musen 2003). We believe that in order to achieve the vision of semantic

CAD/GIS interoperability via Web services and to realize the associated benefits, research in this area needs to go beyond merely merging and mapping ontologies. Techniques must be developed to identify, evaluate, and select the set of ontologies that adequately addresses the given task, and to determine the best order of processing for the set of ontologies to ensure that the intended optimal solution for the task is achieved.

Once a task is decomposed into sub-tasks, a set of available processing ontologies must be identified and evaluated for each domain. The purpose of this evaluation is to select a set of appropriate ontologies that will provide the needed operations. An understanding of various ontologies and their relationship to one another within each domain is critical. Figure 9.4 depicts a possible set of ontologies covering a single domain. While some ontologies are independent of others (i.e., they represent concepts that are not represented by other ontologies within the domain and have no formal relationships), there are others that overlap (e.g., O^1, O^2, and O^3) or have a defined relationship (e.g., O^4, O^6, and O^8). For example, transferring a CAD file to a GIS file may involve multiple ontologies. The oval line on the left hand side of Figure 9.4 represents the set of available ontologies for a domain. However, for any given task only a subset of these ontologies may be needed, and it is possible that some of the needed ontologies will have similar or overlapping scopes that must be determined by mapping pairs of ontologies or by merging two or more ontologies into a single ontology. Figure 9.4 shows an example task where ontologies O^1, O^2, and O^3 are needed, and while O^1 and O^2 are distinct, both overlap with O^3.

In our crane location scenario, one spatial analysis task involves determination of potential obstructions to the crane's movement (e.g., neighboring buildings, surface/sub-surface utilities, roads). Completing this task requires a processing ontology that would capture and describe the specifics of the spatial analysis operations (e.g., inputs, outputs, parameters). If there are multiple ontologies that could provide such knowledge, then the one ensuring the highest semantic confidence

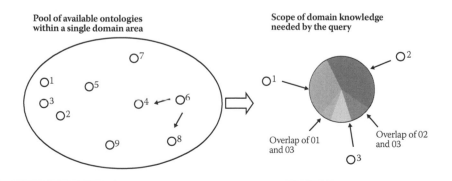

Figure 9.4 **Selection of ontologies from available set of ontologies within a single domain.**

should be selected. Semantic confidence in this context indicates a guarantee that the solution provided by the task is optimal and consistent within the context of the specific application and domain. In some cases, several ontologies may be needed to ensure semantic confidence. For example, addressing the task discussed in our scenario would require both CAD and GIS processing ontologies.

9.4.1.3 Optimal Ontology Processing

Once all the needed ontologies for a given task have been identified, evaluated, and selected, they must be processed in an optimal manner to provide semantic confidence. We model the problem of processing the required ontologies in a graph (see Figure 9.5), with the nodes representing individual ontologies and the links representing the relationship between the ontologies. Each link is assigned a weight $(pw_{x,y})$ which indicates the degree of overlap between pairs of ontologies.

Determining values for $(pw_{x,y})$ requires a technique that evaluates each pair of ontologies for overlapping concepts. The value for $(pw_{x,y})$ may range between 0 and 1 and will represent the amount of overlap between the two ontologies (x,y). A low value indicates a small overlap where a high value indicates a large overlap between the two ontologies. Figure 9.5 depicts those ontologies $(O^2, O^5, O^6, O^7,$ and $O^9)$ that are selected out of the pool of available ontologies in Figure 9.4 and that are needed to address a specific task. The nodes of the graph in Figure 9.5 represent these selected ontologies, and the links represent the relationship, with weights pw, between different ontologies.

With the problem represented as a graph, a solution to semantic confidence is the minimum spanning tree of the graph. In other words, the minimum spanning tree of the graph would guarantee the best solution (utilizing all the needed ontologies with the least amount of overlaps among them) to the task with the highest level of semantic confidence possible. The objective, i.e., determining the minimum spanning tree of the graph using the given weights, is shown in the equation below:

$$\text{Minimize} \sum_{\substack{i=1 \\ j=1}}^{n} (pw)_{ij} \tag{1}$$

where pw is the value assigned to the amount of overlap between the two linked ontologies (i,j). Figure 9.5(b) depicts a hypothetical minimal spanning tree for the graph in Figure 9.5(a). Using this approach, the total amount of overlap is minimized while ensuring that all ontologies (representing the knowledge needed to resolve the entire task) are processed. The minimum spanning tree approach using an algorithm, such as Prim's or Kruskal's (Cormen et al. 2001), will result in the effective utilization of the required ontologies to ensure semantically correct outcomes (or semantics with the highest level of confidence).

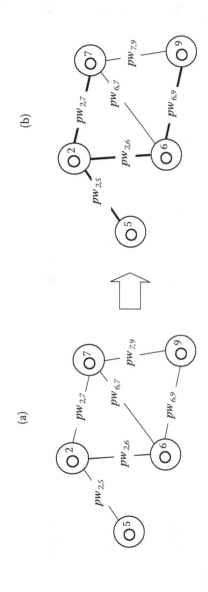

Figure 9.5   A graph representing ontologies and their relationships.

9.4.2 Web Service Matching

Web services are an emerging technology for GIS and CAD applications. As the number of available CAD and GIS Web services increases, finding the suitable ones that provide a solution to the tasks under consideration becomes more difficult. Therefore, effective service discovery is critical to the realization of our approach.

There are methodologies that facilitate service discovery from different perspectives, e.g., Business Process Execution Language for Web Services (BPEL4WS) from IBM, and QoS parameters. Current techniques to discover services in distributed computing environments, such as Universal Description Discovery and Integration (UDDI), have several shortcomings. Examples of these shortcomings are failing to identify the best semantic similarity between service capabilities and user requests, and being unable to ensure that the capabilities of the identified service would meet the needs of the users—such needs are typically described in terms of desired accuracy, price, and response time, among others. For this, there is a need for a new service discovery technique that automatically selects a set of optimal CAD/GIS Web services by satisfying both semantic and QoS criteria for a given task. In other words, such a technique must semantically match between concepts by optimizing QoSs preferred by users. For CAD/GIS integration, research in semantically rich service discovery requires development of algorithms that support publication of new services and service matching.

9.4.2.1 Service Discovery

Existing service matching algorithms usually perform a pair-wise comparison between a service request and all registered services (Li and Horrocks 2004; Paolucci et al. 2002). A match between a request and a service must involve semantic matching of the request concept(s) and the service concept(s). For discovery of CAD/GIS Web services, a service matching algorithm and a QoS filtering algorithm are needed to (1) evaluate semantic similarity between a user request and a registered service; (2) conduct comparisons between user's specific requirements and service(s)' capabilities in terms of QoSs; and (3) provide the user with a list of candidate services for further service composition (see Figure 9.6).

9.4.2.1.1 Service Matching Algorithm

Each sub-task of a given task should be matched to a set of one or more CAD/GIS operations using the AEC-CAD/GIS Mediator and a set of processing ontologies. Information on each operation in these processing ontologies forms the basis of the request for services. The main purpose of the Service Matching Algorithm is to compare the concepts derived from the processing ontology and requirements from the user with the concepts presented in the form of published services. Once a new service profile is generated by the developer, its description is semantically interpreted and registered in an ontology. The matching algorithm will search the relevant ontology to determine the ontological relationship, if any, between the

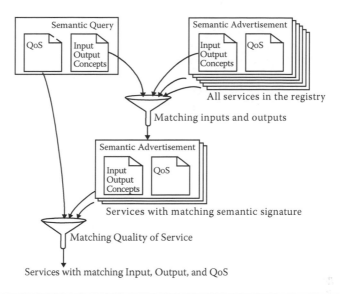

Figure 9.6 Matchmaking procedure.

concepts. Such implicit relationships between service descriptions and requests can be derived through reasoning about the types of match between two output (or input and constraint) concepts as follows. There are four types of similarity match between the output of a service (O^S) and the output of request (O^R), with each matching type assigned a score (Table 9.1).

These definitions of match types are based on previous studies on semantic matching (Li and Horrocks 2004; Paolucci et al. 2002), but utilize a different scoring matrix. Using these definitions of match types, the similarity between a service request and a service description can be reasoned and realized. For example, assuming m concepts in a service description and m corresponding concepts in a service request, the similarity or global match between the request R and the service S can

Table 9.1 Types of Similarity Match Between a Request and a Service

Type	Description
Exact	If O^R and O^S are the same (highest similarity)
Plug-in	If O^R subsumes O^S, then O^S can be used instead of O^R
Subsumption	If O^S subsumes O^R, then the service may not completely satisfy the request
Fail	If O^R and O^S do not have either plug-in or subsumption relations, then the match fails

be derived by summing up the match scores between the concepts pair based on the scoring matrix assigned from domain ontologies (see equations below).

$$\text{match}(C_i^R, C_i^S) \begin{cases} \text{Score1} & \text{if } C_i^R = C_i^S \\ \text{Score1} & \text{if } C_i^R \subset C_i^S \end{cases}$$

$$\qquad \qquad (2)$$

$$\begin{array}{ll} \text{Score 3} & \text{if } C_i^R \supset C_i^S \\ 0 & \text{else} \end{array}$$

$$\text{similarity}(R, S) = \sum_{i=1}^{m} \text{match}(C_i^R, C_i^S) \qquad (3)$$

An ontology is searched for finding the locations of the two concepts, and for determining if they have any relationship using the scoring matrix defined in the equations above. Upon completion of this search and assignment of the scores, each concept will have a record of its sub-classes and super-classes. Then either C_i^R or C_i^S needs to be searched, but not both. For example, Figure 9.7 shows a portion of an ontology, where concept C2 has a super-class C1 and a sub-class C4. Based on the list associated with C2, we can infer that C1 subsumes C2, C2 subsumes C4, and C2 has no relation with C3. To determine if there is a match between a request O^R and a service O^S, the algorithm will search the ontology to find the location of O^R and determine the degree of match between the two concepts by checking to see if O^S appears in O^R's list of sub-classes and super-classes. The same process is followed to match between a service input and a request input. Table 9.2 shows the pseudo-code for the service matching algorithm. In this

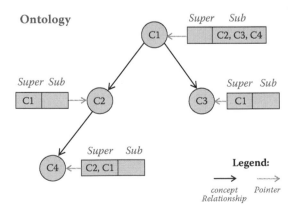

Figure 9.7   Matching concepts in an ontology.

Table 9.2   Pseudo Code for Service Matching Algorithm

Main function match(request, All, G)	//G is an ontology, G = <V, E>
1. var service[ ]	//All is a list of all registered services
2. int match = 0;	// All.length = number of all services
3. **for** i = 1 to All.length **do**	//Find the final score by summing up
4. match = serviceMatch(request, All[i]);	the matching degree
5. **if** match != 0 **then**	
6. add All[i] to service;	//Add scored service to the list of candidate services
7. sort(service);	
8. **return** service;	
function serviceMatch(request, service)	//compare a request with a service
1. int m;	//number of concepts in request and service
2. parse request into concepts c1[m];	
3. parse service into concepts c2[m];	
4. **for** i = 1 to m **do**	
5. u_0 = the root vertex in G;	
6. score[i] = DFS'(u_0, c1[i], c2[i]);	//depth-first search of service concept
7. service.match += score[i];	//calculate match score
8. **return** service.match;	
function DFS'(u, x, y)	//x is request concept
1. **if** u = y **then**	
2. score = degreeOfMatch(y, x);	//y is corresponding service concept
3. **return** score;	
4. **else**	
5. status[u] = "traversed";	
6. **for each** neighbor v of u **do**	
7. **if** status[v] != "traversed" **then**	
8. DFS'(v, x, y);	

(*continued*)

Table 9.2   Pseudo Code for Service Matching Algorithm (Continued)

function degreeOfMatch(u, c)	
1. int score = 0;	
2. **if** c = u **then**	
3. score = "exact" or Score1;	
4. **if** c is subclass of u **then**	
5. score = "plugin" or Score2;	
6. **if** c is superclass of u **then**	
7. score = "subsumption" or Score3;	
8. **return** score;	

matching algorithm, a service request is matched against all the registered services. Whenever a match between the request and any of the registered services is found, the matched service will be assigned a matching score and the services with the highest similarity score will be recorded in the list of candidate services.

A match between a request and a service consists of the match of all the request concepts and the service concepts (function "serviceMatch" in Table 9.2). Here the concepts include all input, output, and descriptions of request and service capabilities. A match is recognized if and only if for each request concept there is a service concept. To determine if there is a match, the algorithm first searches for the service concept in the ontology and then calls the scoring matrix (function "degreeOf-Match" in Table 9.2) to calculate the degree of match (or matching score). The matching scores for all concepts are summed up as the global matching score (or similarity score) between the request and the service. The last part of the algorithm is to sort the resulting matches. The sorting is based on the similarity scores for all matched services. Any sorting algorithm (e.g., insertion sort) can be applied here. After sorting, a list of all candidate services will be provided and will be input to the QoS filtering mechanism.

9.4.2.1.2 QoS Filtering Algorithm

The purpose of the semantic matching algorithm is to select optimal services. Optimal services not only semantically match a user's request, but also best meet the user's preferences (e.g., cost, response time, previous user satisfaction, level of encryption, and accuracy). For this, the QoS filtering algorithm will utilize those candidate services, each with different QoS offerings (which consist of QoS parameters and values), selected by the service matching algorithm. QoS offerings by all candidate services will be checked against user-defined requirements and preferences and the most appropriate ones will be chosen. For the purpose of finding optimal services, a weighting scheme, based on weights for parameters, will be employed. One example would be the level of user satisfaction with candidate Web services, measured on a scale from 1 to 10. Another example would be level of encryption, measured on a scale from 1 to 4.

9.4.2.1.3 Service Discovery Feedback

Once the discovery process is completed, a list of candidate services will be provided. Services with the highest scores would match the user's request with a high semantic confidence. Other identified services may be used as backups based on user flexibility. If an optimal service is not found, the process will search for the next best service. It is reasonable to assume that users will not participate in the development of services, which means the search may result in no services or services that will not satisfy QoS parameters. In such cases, users will be notified and can modify their task or adjust the QoS parameters and rediscover services. When services that satisfy the request cannot be found, they will be marked for future searches to accelerate the service discovery process.

9.4.3 Web Service Composition

Once appropriate services for a given task are obtained, they need to be composed to provide the intended solution. Automated Web service composition is defined as a computerized way of composing a set of available services to accomplish some user-defined task or goal (McIlraith and Son 2002). The concept of planning used in the artificial intelligence domain can be considered as one of the promising techniques for automated Web service composition (Pistore et al. 2004). A number of research efforts have perceived Web service composition as a planning problem (Pistore et al. 2004; Sirin et al. 2004). For instance, Sirin, et al. (2004) leveraged the Hierarchical Task Network (HTN) planning technique along with OWL-S service descriptions (Sirin et al. 2004). Pistor, et al., (2004) viewed Web service composition as planning under uncertainty, where the planning domain is non-deterministic, partially observable, and with extended goals (Pistore et al. 2004). A comprehensive description of the HTN planning technique and planning under uncertainty can be found in Ghallab et al. (2004). We also perceive Web service composition as a planning problem under uncertainty in accordance with Pistor et al. (2004) for the reason described below.

Classical planning techniques rely on such restrictive assumptions as determinism, full observability, and reachability goals. These assumptions are not valid with planning under uncertainty. In a deterministic view, the execution path of each action is fully determined and can therefore be predicted. In a highly dynamic environment like Web services, it is almost impossible to predict everything. This is due to a multitude of sources of uncertainties inherent in a dynamic environment. For example, network latency, availability of services, and quality of available services contribute to uncertainties in Web service composition. In addition to non-determinism, some states of Web service composition may never be observable or observable only after some actions have been executed. For instance, in our example scenario, a planner that composes Web services for transforming CAD data to GIS data cannot know the availability of transformation services until it searches all available CAD/GIS Web services. Further, service composition constitutes a number of sub-goals, and each sub-goal needs to specify requirements of different strengths to take into account non-determinism and possible failure of point of view (Ghallab et al. 2004). In the 3D transform and the data assembly sub-goal, there may be either one single service or a set of multiple services (e.g., one individual service for co-ordinate transformation, workspace generation, and workspace aggregation). Thus, from the non-determinism and possible failure point of view, it is safe to invoke one single service to achieve 3D transformation and the data assembly sub-goal instead of invoking multiple services. Thus, Web service composition has the characteristics of a problem involving planning under uncertainty.

Planning under uncertainty has been extensively studied in the domain of robotics, manufacturing, and logistics. Popular approaches for solving planning under uncertainty have been focused on leveraging the Markov Decision Processes (MDP), Partially Observable Markov Decision Processes (POMDP), and planning under model checking (Ghallab et al. 2004; Russell and Norvig 2003; Thrun 2002). With the MDP approach, the key idea is to formalize a planning problem as an optimization problem. Uncertainty related to action outcomes is modeled with some kind of probability distribution function (Ghallab et al. 2004). The goals are represented using utility functions, which are numeric functions giving preferences to actions to be executed. There are a number of viable plans, and such plans as policies that specify the action to perform in each state. The objective of this approach is to search for a plan that maximizes the utility function. The difference between MDP and POMDP is that POMDP can handle partially observable states. An alternative to the MDP approach is a planning approach based on Model Checking. The main idea is to solve the problem model theoretically where sets of states and transitions are represented and manipulated symbolically (Ghallab et al. 2004). Such symbolic representation and manipulation often result in compact representation, thus saving computational time.

The aforementioned algorithms have been primarily tested on robotic applications (e.g., robot navigation), assembly, and manufacturing. Limited research studies have been done on leveraging them for Web service composition (Ghallab et al. 2004; Pistore et al. 2004). Current research studies have focused on the composition of a limited number of services (Pistore et al. 2004). Hence, it is still not clear whether the existing algorithms can scale well (in terms of computational time and resources) when the number of available services is vast (which is typical of Web services). The MDP and POMDP algorithms seldom scale up when the number of states increases. The current success in Web service composition using MDP/POMDP is based on a simple case scenario with a few services (Doshi et al. 2004). Therefore, it is necessary to understand the applicability and scalability of existing algorithms for the Web service composition problem.

Further, MDP and POMDP leverage the concept of probabilities, utility functions, and optimization to solve uncertainty-based planning problems. Such difficulties as calculation of utility function and probabilities have not been addressed yet. There is not enough statistical data to express state transitions (in MDP and POMDP) in terms of probabilities. In addition, defining a utility function in terms of QoS and assigning costs to these different QoSs have not been explored yet. Thus, there is a need for further exploration and formalization of the abovementioned problems.

9.5 Conclusion and Future Research

The lack of interoperability between CAD and GIS platforms results in inefficiency and increased costs. While current community and vendor efforts to address this issue have resulted in a low level of integration between existing platforms, significant potential benefits are still possible by enabling interoperability at the semantic level. We envision semantic CAD/GIS Web services as a means for solving problems requiring both CAD and GIS data and operations, and for insulating users from gaps in knowledge and ambiguity in semantics by allowing them to focus on addressing domain tasks. Such a semantic Web service approach consists of task interpretation, Web service matching, and Web service composition. For task interpretation, there is a need for development of key algorithms and associated metrics to identify and evaluate ontologies needed to provide the required knowledge to solve cross-domain problems. In the area of Web services, algorithms to support the publication of new services and perform service matching in support of CAD/GIS integration must be developed. Additionally, research is required to identify and quantify QoS parameters, and to develop a mechanism to provide feedback on the results of service discovery to the user in the event that the desired services cannot be identified. Finally, while Web service composition can be characterized as a planning under uncertainty problem, the inherent differences between Web service composition and existing planning under uncertainty problems pose additional research questions with respect to the applicability and scalability of existing algorithms.

9.6 Acknowledgments

This work is supported in part by the Pennsylvania Infrastructure Technology Alliance (PITA), a partnership of Carnegie Mellon University, Lehigh University, and the Commonwealth of Pennsylvania Department of Community and Economic Development. We also acknowledge Kristen Bailey and Sharad Oberoi for their help in the earlier parts of this research.

References

Akinci B., Tantisevi K., and Ergen E. "Assessment of the capabilities of a commercial 4D CAD system to visualize equipment space requirements on construction sites." *Construction Research 2003*. Honolulu, Hawaii, p. 116.

Ambriola V. and Gervasi V. (2006). "On the systematic analysis of natural language requirements with CIRCE." *Automated Software Engineering* 13(1): 107–167.

Autodesk. (2007). "AutoCAD." http://www.autodesk.com/ (accessed August 10, 2007).

BARBi. (2007). "Building and construction reference data library." http://www.barbi.no/index.jsp (accessed August 10, 2007).

Boury-Brisset A. C. (2003). "Ontology-based approach for information fusion." *Information Fusion 2003*, vol. 1: 522–529.

Chen P. P. S. (1997). "English, Chinese and ER diagrams." *Data and Knowledge Engineering* 23: 5–16.

Cormen T. H., Leiserson C. E., Rivest R. L. and Stein C. (2001). *Introduction to algorithms*, The MIT Press, Cambridge, Massachusetts.

Cruz I. F., Sunna W., and Chaudhry A. (2004). "Semi-automatic ontology alignment for geospatial data integration." *Geographic Information Science*, 51–66.

Doshi P., Goodwin R., Akkiraju R., and Verma K. "Dynamic workflow composition using Markov decision processes." *Proceedings of the IEEE International Conference on Web Services (ICWS'04)*, 576–582.

e-Cognos. (2007). "e-COGNOS Ontology " http://e-cognos.cstb.fr/Downloads/Download.htm (accessed June 12, 2007).

El-Diraby T. A., Lima C., and Feis B. (2005). "Domain taxonomy for construction concepts: Toward a formal ontology for construction knowledge." *Journal of Computing in Civil Engineering* 19(4): 394–406.

El-Diraby T. E. and Kashif K. F. (2005). "Distributed ontology architecture for knowledge management in highway construction." *Journal of Construction Engineering and Management* 131(5): 591–603.

ESRI. (2007). "ArcGIS Desktop." http://www.esri.com/software/arcgis/about/desktop_gis.html (accessed August 10, 2007).

Fensel D. (2002). "Ontology-based knowledge management." *Computer*, 35(11): 56–59.

Fensel D., van Harmelen F., Horrocks, I., McGuinness D. L., and Patel-Schneider P. F. (2001). "OIL: An ontology infrastructure for the Semantic Web." *Intelligent Systems, IEEE [see also IEEE Intelligent Systems and Their Applications]*, 16(2): 38–45.

Fonseca F., Davis C., and Camara G. (2003). "Bridging ontologies and conceptual schemas in geographic information integration." *GeoInformatica* 7: 355–378.

Fonseca F. T., Egenhofer M. J., Agouris P., and Mara G. (2002). "Using ontologies for integrated geographic information systems." *Transactions in GIS* 6: 231–257.

Ghallab M., Nau D. and Traverso P. (2004). *Automated Planning: Theory & Practice*, Morgan San Francisco, California: Kaufmann Publishers Inc.

Gruber T. (1993). "A translation approach to portable ontology specifications." *Knowledge Acquisition*, 5(2): 199–220.

Guo S.-J. (2002). "Identification and resolution of work space conflicts in building construction." *Journal of Construction Engineering and Management* 128(4): 287–295.

Harmain H. M. and Gaizauskas R. "CM-Builder: An automated NL-based CASE tool." *Automated Software Engineering, 2000. Proceedings ASE 2000. The Fifteenth IEEE International Conference on*, 45–53.

IAI. (2007). "IFC/ifcXML Specifications." http://www.iai-international.org/Model/IFC (ifcXML)Specs.html (accessed August 10, 2007).

Jones S. A. (2005). "Interoperability issues for underground utilities: a case study of the NASA Langley research center." *Public Works Managements Policy* 9(3): 232–247.

Karimi H. A., Akinci B., Boukamp F., and Peachavanish R. "Semantic interoperability in infrastructure systems." *Information Technology 2003*. Nashville, Tennessee, p. 42.

Klien E., Lutz M., and Kuhn W. (2006). "Ontology-based discovery of geographic information services—An application in disaster management." *Computers, Environment and Urban Systems* 30(1): 102–123.

Kotis K., Vouros G. A., and Stergiou K. (2006). "Towards automatic merging of domain ontologies: The HCONE-merge approach." *Web Semantics: Science, Services and Agents on the World Wide Web* 4(1): 60–79.

Li L. and Horrocks I. (2004). "A software framework for matchmaking based on semantic Web technology." *International Journal of Electronic Commerce* 8(4): 39–60.

Mark D., Smith B., and Tversky B. (1999). "Ontology and geographic objects: An empirical study of cognitive categorization." *Lecture Notes in Computer Science*, 1661, 283–298.

McIlraith S. and Son T. "Adapting Golog for composition of semantic Web services." *Proceedings of the Eighth International Conference on Knowledge Representation and Reasoning (KR2002)*, 482–493.

Mich L. (1996). "NL-OOPS: From natural language to object oriented requirements using the natural language processing system LOLITA." *Natural Language Engineering* 2(2): 161–187.

Noy N. F. and Musen M. A. (2003). "The PROMPT suite: Interactive tools for ontology merging and mapping." *International Journal of Human-Computer Studies* 59, 983–1024.

OGC. (2007). "Open Geospatial Consortium, Inc." http://www.opengeospatial.org/ (accessed August 10, 2007).

Paolucci M., Kawamura T., Payne T. R., and Sycara K. (2002). "Semantic matching of Web services capabilities." *The Semantic Web—ISWC 2002: First International Semantic Web Conference, Sardinia, Italy, June 9–12, 2002. Proceedings*, p. 333.

Peachavanish R. and Karimi H. A. (2007). "Ontological engineering for interpreting geospatial queries." *Transactions in GIS*, 11, 115–130.

Peachavanish R., Karimi H. A., Akinci B., and Boukamp F. (2006). "An ontological engineering approach for integrating CAD and GIS in support of infrastructure management." *Advanced Engineering Informatics* 20(1): 71–88.

Pistore M., Barbon F., Bertoli P., Shaparau D., and Traverso P. (2004). "Planning and monitoring Web service composition." *Artificial Intelligence: Methodology, Systems, and Applications*, 106–115.

Rasdorf W., Shuller E., Poole R., Abudayyeh O., and Robson F. (2000). "Information management at state highway departments: Issues and needs." *ASCE*, 134–142.

Russell S. J. and Norvig P. (2003). *Artificial Intelligence: A modern approach*, Upper Saddle River, New Jersey: Pearson Education.

Sirin E., Parsia B., Wu D., Hendler J., and Nau D. (2004). "HTN planning for Web service composition using SHOP." *Journal of Web Semantics*, 1, 377–396.

Smith B. and Mark D. M. (2001). "Geographical categories: An ontological investigation." *International Journal of Geographical Information Science* 15, 591–612.

Tantisevi K. and Akinci B. (2007). "Automated generation of workspace requirements of mobile crane operations to support conflict detection." *Automation in Construction* 16(3): 262–276.

Thrun S. (2002). "Probabilistic robotics." *Communications of the ACM* 45(3): 52–57.

Varghese K. and O'Connor J. T. (1995). "Routing large vehicles on industrial construction sites." *Journal of Construction Engineering and Management* 121(1): 1–12.

W3C. (2007). "Technical Reports." http://www.w3.org/TR/ (accessed August 10, 2007).

Index

223

Printed and bound by CPI Group (UK) Ltd, Croydon, CR0 4YY

23/10/2024

01777670-0007